THE BIOCYCLE GUIDE TO COMPOSTING MUNICIPAL WASTES

EDITED BY THE STAFF OF

BIOCYCLE

JOURNAL OF WASTE RECYCLING

The JG Press, Inc.

The JG Press, Inc.
Box 351
Emmaus, Pennsylvania

Printed on recycled paper in the United States of America.
ISBN 0-932424-08-2
Second Printing

CONTENTS

SECTION ONE
COMPOSTING RATIONALE AND ACTIVITY

SECTION TWO
DESIGN AND OPERATIONS

SECTION THREE
END PRODUCT STABILITY AND MATURITY

SECTION FOUR
HEALTH ISSUES AND COMPOST SAFETY

SECTION FIVE
MARKETS AND UTILIZATION

SECTION SIX

CO-COMPOSTING OF SOLID WASTES

SECTION SEVEN

LEAF AND YARD WASTES

SECTION EIGHT

INDUSTRIAL AND HAZARDOUS WASTES

SECTION NINE

THEORY, PRACTICE AND SUMMARY

APPENDIX

DIRECTORY

SECTION ONE

COMPOSTING RATIONALE AND ACTIVITY

1

THE RATIONALE FOR COMPOSTING

Clarence Golueke

A SURVEY of past and present developments in the composting of municipal wastes reveals an overall trend that probably can best be described as taking the form of a tightly coiled horizontal spiral. The linear advance made by the spiral reflects the overall progress made in attitudes, approaches and technology. The cyclical element pertains to the rise, fall, and subsequent resurgence of particular attitudes, approaches, and technological developments. The qualification "tightly coiled" is used to reflect the slightness of the advance made at each cycle.

Protection of public health, conservation of resources (plant nutrients, organic matter), safe disposal of wastes, reduction of environmental stress, and attainment of favorable economics have always been named as key motives (i.e. moving forces) for composting. The order of their priority, however, in the selection of composting in preference to other waste management options has varied with the passage of time. To be completely objective and realistic, one must admit that economics has always constituted a central, pervasive theme about which all other motives revolved and continue to do so today.

In the early 1900's, open dumping of refuse was the norm. Because sewage treatment was minimal and sparsely practiced, the volumes of sludge produced were correspondingly small. Any motivation toward composting was tied to resource conservation and the protection of public health.

In time (mid-1900's), as the citizenry became aware of the environmental affront and the unfavorable public health significance of the open dump, it began to demand the adoption of a more enlightened and effective way of managing municipal wastes. This new attitude, combined with the rapid increase in the rate of waste generation, forced development and utilization of alternatives. Landfilling began to lose some of its attraction due to the dawning realization that land was rapidly becoming either unavailable or unreasonably expensive, or both. In addition, incineration was expensive, and its emissions were becoming the focus of public health and environmental concern. The outcome was a renewal of serious consideration of composting as a suitable means of waste disposal, and so waste disposal became a leading motive for composting.

With the coming of the mid-1970's, resource conservation again reached top priority, which it continues to share with protection of public health and the imposition of lesser environmental stress.

The order of priorities was shifted by the problem of what to do about the increasing output of wastewater sludges that accompanied the advances made in sewage treatment. This factor began to assume importance in the mid 1970's and continues to be a prime moving force today. Because the nature of sludge lends itself to a compost technology that is relatively simple, the costs involved in sludge composting are correspondingly moderate, making this management option especially attractive. The favorable economics depends in part on the fact that through composting, valuable plant nutrients and organic matter in the sludges are recovered and conserved and simultaneously rendered safe to use. These additional factors join economics in being the primary motives for composting sewage sludge.

Interest in sludge composting rose during the 1970's, with composting of solid wastes all but vanishing. The 1980's, however, have brought a resurgence of interest in solid waste composting as a management option. This revival has been directly linked to skyrocketing landfill costs and diminishing landfill space, increased opposition to incineration, and favorable economics. The success of sludge composting has aided this rising interest, as well as kept the general field of composting moving along a linear path.

The Biological Advantage

The ultimate goal of composting should be the promotion of our mental and physical well-being. Composting contributes to the attainment of this well-being by providing a means of: (1) Safeguarding public health; (2) Preserving or improving the quality of the environment; (3) Conserving resources; and (4) Enjoying the advantages peculiar to biological systems of treatment. Throughout the pages of this text, these issues will be addressed.

The statement, "enjoyment of biological systems," pertains to composting as a waste treatment process. Waste treatment can be accomplished by physical means, chemical means, biological means, or a combination of two or all three of those means. Until relatively recently, and possibly even at present, biological treatment systems have been the types most commonly used. This is especially so in wastewater treatment. With municipal solid wastes, it usually was in the form of the decomposition that takes place in landfills, and less commonly as composting. The advantages of biological over physical or chemical systems in waste treatment are:

(1) They are more compatible with a high quality environment.
(2) They are simpler to operate.
(3) They require less expensive equipment and operation.

The compatibility with a good environment comes from the fact that the processes are natural. The second and third advantages are the consequences of biological systems being natural processes, and thus lack the need for precise control and sophisticated equipment. Of course, biological systems, like all other processes, have their disadvantages, chief of which is slowness. With physical and chemical processes, one designs in terms of holding times of minutes or at the most an hour or two; with biological systems, the range is from several hours to months.

Principles

Probably all of the principles involved in composting wastes can be traced to the single all-inclusive fact that composting is a biological process. This sounds like a simple statement, but after a little thought, its broad ramifications become apparent. Two broad inferences can be drawn from this fundamental: (1) Composting has the limitations of biological systems; and (2) The process is affected by the basic environmental conditions that influence all biological activity.

Definition

Simply stated, composting is the biological decomposition of the organic constituents of wastes under controlled conditions. The term "decomposition" is used instead of "stabilization," because when applied in practical usage, the process rarely is carried on to the point at which the waste is completely stabilized. In fact, the cut-off point may at times be short of "temporary stabilization." The term "biological" distinguishes composting from other types of decomposition, such as chemical or physical (e.g., incineration, pyrolysis, etc.). "Organic" is applicable because, with few exceptions, only the organic portion of wastes is subject to biological breakdown. A very important term in the definition is "controlled." It is the application of control that distinguishes composting from the natural rotting, putrefaction, or other decomposition, that takes place in an open dump, a sanitary landfill, in a manure heap, in an open field, etc.

Classification

Compost systems can be classified on three general bases: oxygen usage, temperature, and technological approach. Oxygen usage is divided into aerobic and anaerobic. When temperature serves as the basis, the division becomes mesophilic and thermophilic. Finally, using technology as the key, the classification is divided into static pile or windrow, and mechanical or "enclosed" composting. The terms in each of the classifications are practically self-explanatory.

Aerobic composting involves the activity of aerobic microbes, and hence the provision of oxygen during the composting process. The opposite prevails in anaerobic composting—i.e., anaerobic bacteria accomplish the decomposition and oxygen (air) is excluded from the composting mass. Aerobic composting generally is characterized by high temperatures, the absence of foul odors, and is more rapid than anaerobic composting. Anaerobic decomposition is characterized by low temperatures (unless heat is applied from an external source), the production of odorous intermediate (reduced) products, and generally proceeds at a slower rate than does aerobic composting. The advantages of aerobic composting over anaerobic composting arise from the differences in their characteristics—in other words, aerobic composting is more rapid, permits high temperatures, and is not odorous. The main advantage in anaerobic composting is that the process can be carried on with a minimum of attention, and as such can be sealed from the environment. Because of its many advantages, most modern compost processes are basically aerobic—or attempt to be.

As the term implies, in mesophilic composting the temperatures are kept at intermediate temperatures (15° to 40°C), which in most cases is the ambient temperature.

Thermophilic composting is conducted at temperatures from 45°C to 65°C. In practice, most processes include the two ranges.

Efficiency and Performance

The efficiency and performance of a biologically-based system, however technologically advanced it may be, are directly dependent upon the well-being of the microorganisms upon which the system is based.

It is the activity of these microorganisms that constitutes the driving force of the biological process for which the system is designed or intended to promote. The growth and activity of the microorganisms, in turn, depends upon the extent to which all biologically essential factors are present.

Consequently, the factor in lowest supply or exhausted most rapidly becomes the factor that limits the growth and activity of the organisms under consideration, and hence is appropriately termed the "limiting factor."

A corollary of the principle of the limiting factor is that unless each and every factor is adequately met, problems will arise. Obviously then, problem prevention becomes a matter of designing and operating a given system such that no operational, nutritional, environmental, or any other conceivable factor be deficient, i.e. becomes limiting. Yet another facet of the principle of limiting factor is its positive relation to sound economics.

Fads and Moderation

With respect to compost technology and practice, the theme "fads and moderation" parallels the principle of the limiting factor in a number of ways. One parallel is the fact that the majority of the fads that appeared in the past several decades of compost practice were characterized by an overemphasis of one or two environmental or operational factors and a neglect of other factors.

The rather dismal record of problems, reverses, and failures, characteristic of many compost operations during the past decades, originated in the combination of overemphasis and neglect associated with the fads current at the times of the operations. Fads generally pertain to systems that usually are fully or partially mechanized and are in-vessel or sheltered, and are currently touted as being better than any other existing system. The past record of compost practice is a reminder that all too often fads ignore the fact that each and every biological activity proceeds at the pace and manner determined by the factor in least "abundance," and independently of the system.

The solution, of course, is moderation. Moderation implies that attention be given to all factors and that no factor be neglected. Therefore, by giving due obeisance to moderation, many of the problems associated with fads can be avoided, at least the problems of microbial origin.

Essential Factors

Briefly stated, essential factors are those features of the physical, chemical, and biological background that are necessary to the establishment and proliferation of the microorganisms specific to a desired process. The remainder of this chapter deals with five essential factors that have become key design features in recent compost technology, namely, suitable microbial population or populations, aeration (oxygen availability), temperature, moisture content, and carbon availability.

Microbial Population

Type and size of microbial population rarely constitute a limiting factor in the composting of most crop residues, manures, and assorted agricultural wastes, and in the composting of sewage (domestic) sludges and the nontoxic organic components of municipal solid waste (MSW). Indeed, with respect to such materials, the major shortcoming regarding the microbial population factor is likely to be in terms of superfluity rather than of scarcity.

It should be emphasized that there are exceptions to the superfluity of inoculums in composting. Thus, microbial population could become the limiting factor in the composting of residues or wastes with these characteristics: 1) are almost completely homogeneous in composition; 2) have been sterilized; 3) are residues from the production of antibiotics (e.g. pharmaceutical wastes); 4) have biologically degradable toxic organic compounds ("hazardous wastes"); or 5), are a combination of two or more of the preceding four.

An important point to remember about these exceptions is that in such cases, microbial population rarely is the sole candidate for the position of limiting factor. One or another nutritional factor usually is also in short supply (e.g. nitrogen, phosphorus). Identifying and supplying microbes appropriate to these special situations are the objectives of a variety of research and pilot projects now being actively pursued, especially with respect to petroleum refinery residues, toxic refractory organics, and pesticides.

Aeration/Oxygen Availability

Aeration, i.e. aerobiosis, was not always held in the high esteem accorded it today. From the days of Sir Albert Howard and his "Indore" system of composting until the mid-1950s, the relative virtues of anaerobic composting and aerobic composting were vigorously debated. In reality, the debate was largely academic in that complete aerobiosis could not have been attained in a composting mass with the systems available at that time. For example, the Indore process was considered as being an aerobic system, despite the fact that the composting mass was contained in a trench and was turned only once or twice each month.

Probably, the present exaggerated opinion of the importance of aeration is a reaction traceable to the difficulty, if not impossibility, of providing adequate aeration with the means available in the early days. In fact, a hasty perusal of early reports (1950s, early 1960s) reveals a tendency to regard aeration as the cure-all for all compost problems.

With the arrival of the 1960s, increasingly convincing arguments spawned by advances in knowledge moved the consensus to favor aerobic composting. The swing was irreversibly accelerated by the inescapable fact that compost operations involving anaerobic systems were particulary noteworthy for the pervasive foul odors in their immediate and not-so-immediate vicinity.

Unfortunately, because of the modern penchant for simplistic solutions, the usual procedure in system design is to concentrate on the maintenance of aerobiosis in excess of that required for an efficient and nuisance-free compost process—in other words, aerate to excess. Excessive aeration is not necessarily detrimental, except that at times it may not be compatible with the maintenance of an optimum temperature. Another undesired consequence is that excessive aeration may promote evaporation to a point at which moisture becomes limiting.

Temperature

As with aerobiosis vs. anaerobiosis, the dispute over the relative virtues and drawbacks of thermophilic composting vs. mesophilic composting began almost simultaneously with the first efforts at "formalizing" composting, i.e. the establishment of procedures and standards and the defining of applications. For example, common items in the early compost lore were references to the unfavorable effect of excessive temperatures and the uncertainty of many of the benefits attributed to high temperatures. It also became fairly widely known that spore forming microbes passed into the spore stage and nonspore forming obligate mesophylls disappeared at temperatures above 60°C.

Nevertheless, from early on the more common conviction was that satisfactory composting and the attainment of high temperatures are mutually interdependent. This favorable opinion of high temperature composting was strengthened by the fact that sanitation specialists suspected and then demonstrated that high compost temperatures led to the destruction of pathogenic organisms. Hence, from its earliest days, composting was recognized as being an effective means of sanitizing wastes.

With the passage of time and coincidental increase in exposure to physics and chemistry and attendant principles, especially the laws of thermodynamics, biologists began to attribute more and more advantages to thermophilic operations. Indeed, the attainment of high temperatures eventually became a goal for two main reasons: 1) promotion of process efficiency and effectiveness (accelerate the process), and 2) destruction of harmful organisms.

But as so often happens, excess began to creep in, and high temperature became a goal in itself, and its relation to and bearing on other factors receded into the background. Furthermore, the hygienic effectiveness of high pile temperatures began to be overestimated. Thus, the mere attainment of high temperatures in a pile was regarded and continues to be regarded as constituting unequivocal assurance that the composted material is hygienically safe. Fortunately, through research efforts (discussed later), the pendulum has begun to swing to a more realistic view of high temperature and its advantages and disadvantages.

Despite all that has been said in the preceding paragraphs about thermophilic temperatures, it must be admitted, albeit reluctantly, that excluding its relation to sanitation, the position (if it has one) of high temperature as a limiting factor is as yet to be determined. On the other hand, the limiting effect of cryophilic temperature is well established.

Moisture Content

A recent development that illustrates the close parallel between the two subjects, "fads and moderation" and "limiting factors and problem prevention," is the inordinate emphasis on the production of a dry compost product. An outcome of the emphasis is the mistaken equating of dryness with true stability.

Although it is true that a dry product may seem to be stable, such stability is transient and immediately vanishes when moisture ceases to be a limiting factor. On the contrary, because moisture can easily become the limiting factor in a composting operation and true stability (the proper goal of composting) is an intrinsic characteristic reached only through biological decomposition, dryness of product should be a consideration only in the final conditioning of the product.

Apparently, the rationale for the inordinate emphasis on dryness of product is twofold: 1) a material is easier to manipulate when dry than when wet; and 2) if kept dry, it can be stored without becoming a nuisance. This rationale is truly valid in its proper place, i.e. as applied to the conditioning of the product at the completion of the compost process. That being true, drying becomes a logical step in the conditioning of the product after it has been composted and as a prelude to its storage and marketing or utilization.

Carbon and C/N Ratio

Brief mention is made of an essential nutrient factor that is receiving increasing attention because of its bearing on the composting of sewage sludge, namely carbon as it relates to the carbon:nitrogen (C/N) ratio. The problem has not been so much the excessive emphasis on it, but rather the erroneous assumption that the carbon in the woodchips and sawdust used as bulking agents is readily available to bacteria.

The fact is that with the exception of green (uncured) wood (e.g. sapwood), most of the carbon in wood becomes available to bacteria at a very slow rate. The reason is that practically all of the carbon in the two materials is in the form of xylem ("wood" and "xylem" are synonymous in arboreal terminology). For the most part, xylem is lignaceous in composition, and lignin is highly resistant to microbial attack. Among the exceptions are the bacteria in the termite gut and some fungi. Green (uncured) wood and sapwood may contain some wood sugars that are more or less readily susceptible to bacterial attack.

The upshot is that for practical purposes, woodchips and sawdust should be considered almost entirely as bulking and moisture absorption agents, and very little, if at all, as sources of carbon.

Bearing of Limiting Factors on Good Economics

The relation between limiting factors and good economics is quite straightforward. If a factor becomes limiting, the compost process is slowed and problems develop (e.g. insufficiency of oxygen and accompanying generation of foul odors). A slowing of the process results in inefficient use of equipment and in added space and labor requirements, and less product to use or market.

All consequences of process slowing translate into monetary costs. Solving problems, even the trivial, involves the expenditure of money that would not be spent had the problems not developed.

In all compost operations, paying careful attention to biological, operational, and environmental factors is good practice in terms of optimum performance, avoidance of problems, and attainment of sound economics.

2

COMPOSTING ACTIVITY: MUNICIPAL SLUDGE

THE level of composting activity has increased dramatically in the 1980's, not only of municipal sewage sludge, but of leaves and yard wastes, industrial wastes, some hazardous materials, and increasingly, the organic and heavy fractions of the municipal solid waste stream. This chapter presents an overview of this surge in municipal sludge composting activity in the United States and in other countries.

BioCycle Magazine conducts an annual survey of sewage sludge composting facilities in the United States. The first full-scale survey ever taken was done in 1983 by George Willson and David Dalmat. That survey found 61 facilities in operation, and 29 in the planning, construction or design phase—for a total of 90.

Since then, the numbers have changed dramatically. The 1988 data shows that there are 114 operating sludge composting facilities, and 105 in the consideration, planning, design or construction phase—for a total of 219. (see Table I for details.)

The four most common methods for composting sewage sludge are aerated static pile, windrow, aerated windrow, and in-vessel. In addition, one facility is proceeding with vermicomposting, using earthworms to further process the compost material.

The number of in-vessel systems in operation has increased dramatically. In 1985, there were three plants in operation; in 1986 there were eight and in 1987, there were 15. This year, there are 19 in operation, four under construction, and 20 in the planning, design, or bid phase. The aerated static pile method is still the overall leader, with 61 in operation, 10 in construction, and 13 in the planning, design or bid phase. There are 29 operating windrow facilities, and three in the planning, design or bid phase. (Interestingly, there are seven pilot projects experimenting with the windrow method.) There are five operating aerated windrow facilities, and three in the planning, design or bid phase.

Most municipalities and counties are composting a dewatered or sand bed dried sludge. Belt filter presses appear to be the most common type of dewatering system used.

The general climate for sludge composting is healthy. More cities are turning to composting because of the public's positive perception of recycling. And as it gets increasingly difficult to site combustion facilities—and where land application is not feasible—composting is becoming the preferred method for handling sewage sludge.

Increasingly, landfill closures and sky-rocketing landfill tipping fees are forcing wastewater treatment plants to evaluate and choose other sludge management options. State policies are also either eliminating or effectively discouraging that option. New Jersey, for example, has banned sludge disposal in landfills. Rhode Island has ordered a landfill in the state that accepts about 70 percent of all sludge generated in Rhode Island to reduce that amount by 75 percent by late 1989. Elsewhere, many landfill operators are no longer accepting sludge. Composting is being considered or chosen by manicipalities or counties affected by the landfill crisis.

In many states, however, land application is still the preferable and least costly method of disposal. Composting, in some cases, is actually the back-up method to land application projects, used when haulers cannot get onto agricultural land. Elsewhere, shifts are being made from land application to composting because land development is forcing municipalities to haul further and further away, drastically changing the economics of that disposal method.

The rising interest in composting is coinciding—and in part fueled by—the increasing sophistication in the technology and the level of operational knowledge. Plant operators, city officials and consulting engineers are able to take a more detailed look at the options, and are more aware that compost systems are not black boxes.

That level of sophistication applies to the higher "tech" in-vessel systems as well as the lower "tech" alternatives. Some aerated static pile facilities, for example, have sophisticated materials handling and odor control systems. A number of operations—static pile, windrow and aerated windrow—are enclosing many parts of the operation to enhance both odor and process control.

Research on a variety of aspects of sludge composting is being conducted at operating facilities. Odors are the number one problem experienced by composting operations, and controlling them is the primary target of research dollars. Denver, Colorado, Akron, Ohio, Montgomery County, Maryland, and Schenectady, New York are just a few of the sites with ongoing odor control studies.

In addition to odors, operations—particularly aerated static pile—cited problems with moisture control. High moisture contents make it difficult to get the compost dry enough to screen. This problem is acute in regions with cold and rainy weather. Research is also being conducted on aeration, mixing and product quality. Experimentation is being done with bulking agents to control odors and moisture.

For example, Scarborough, Maine began its aerated static pile composting operation (30 cu.yd./wk.) utilizing wood chips as a bulking agent. After a few months of operation, the facility was faced with space constraints for pile construction and screening, and experiencing odor problems. The operators decided to experiment with bio-ash from a nearby paper company, and found that this bulking agent led to a reduction in the odors. In addition, less space was required (the 1:1 wood chip; sludge mixture would cover three to four pipes whereas the bio ash/sludge mixture covers one pipe) and there wasn't any need to screen. Scarborough now uses only wood ash in its operation.

While wood chips and sawdust are the two most common bulking agents/amendments, facilities have been trying a variety of materials, such as the following: rice hulls and coffee grounds, leaves and yard waste, feedlot manure, and rubber tires.

One innovation we expect to see more of in the future is the composting of sludge with leaves, yard waste, tree trimmings, and the heavy organic fraction of the solid waste stream. Several facilities are already co-composting with these materials, while a few others have such plans in the works.

O&M Costs

The per dry ton operating costs for sludge composting operations varied extensively. For aerated static pile facilities, the cost/dry ton ranged from $59 to $300, with the majority falling between $125 and $175. The average cost (based on data from nine facilities) was $154/dry ton. Three operating windrow composting projects reported costs of $80.50, $152 and $158, with an average of $130/dry ton. The lowest cost/dry ton for in-vessel composting reported was $71.19, and the high was $325. The average cost (based on data from five facilities) was $175/dry ton. One aerated windrow facility reported costs of $100/dry ton.

The Market Report

Several years ago, uncertainty about markets was cited as one of the primary drawbacks to composting. That concern is rarely voiced anymore.

Of the 43 facilities providing marketing information, 26 are or plan to market the compost through the public works agency and 12 sell or plan to sell the compost to a private company, which in turn markets the material. Five facilities use a combination.

The two leading markets for sludge composts, based on data from the surveys, are landscaping/general contracting and public works projects (including public parks). Nurseries are also dominant users, as are homeowners. Six of the facilities reporting have giveaway programs to the general public. Other markets cited for the compost include agricultural applications, sod farms, topsoil manufacturers, golf courses, retail garden shops, and tree farms. The compost also is used for landfill cover and strip mine reclamation. Only one facility reported landfilling the material.

Data from the surveys shows that the price of compost ranges from $1/cubic yard to $20/cubic yard, with the following breakdown: 10 facilities in the $1-$5/cubic yard range; nine in the $5-$10/cubic yard range; and two that charge over $10/cubic yard. Four facilities reported selling compost by the bag as well, with prices ranging from $2 to $3.75/bag.

Trends For The 1990s

From the few pioneers of the early 1970s, to well over 100 facilities operating today, composting has evolved into a sound method for managing municipal sewage sludge. Operating facilities not only successfully process sludge and distribute the compost, but act as ongoing research centers, developing effective odor control systems, testing a variety of bulking agents, experimenting with aeration rates and pressures, and finetuning materials handling equipment and methods.

The positive experience with sludge composting is carrying over to other waste streams such as MSW and industrial and agricultural byproducts. There is a gradual trend toward co-composting, particularly with leaves and yard waste and the heavy organic fraction of the solid waste stream.

Another trend that is growing stronger is the concept of a regional composting facility, where a number of communities join together to implement a project. This approach is being discussed by a number of towns in the Northeastern states.

There is concern among compost facility operators, as well as state sludge management coordinators, about the potential of the Environmental Protection Agency's pending technical regulations for sludge management severely restricting the distribution of sludge composts. While some states are rewriting their regulations to open up markets for sludge composts, it is possible that the EPA regulations may impose limits that would have the opposite effect.

There is a growing trend among states to develop standards and specifications for sludge composts (as well as other types of composts, e.g. MSW, leaf and yard waste, etc.). Some states have issued regulations that get into operational aspects of facilities. New Jersey for example, issued a sluge management guideline that requires all future composting facilities in the state to be enclosed—either in-vessel systems or enclosed static piles or windrows. Currently operating facilities have been grandfathered in, although any expansion of those plants must be enclosed. The objective of the regulation, according to a state sludge management official, is to improve odor control.

SLUDGE COMPOSTING FACILITIES IN THE UNITED STATES

STATE PLANT NAME	STATUS	TYPE	SLUDGE VOLUME *dry ton/day (unless noted)*
1. ALABAMA			
1. Dothan City	Operational	In-Vessel (Taulman)	4
2. ALASKA			
1. Homer	Design (complete)	A-SP	750/year
2. Juneau	Consideration		3-4
3. ARIZONA			
1. Phoenix: 23rd Ave. Plant	Operational (by Western Agricultural Products)	Windrow	
2. Phoenix: 91st Ave. Plant	Operational	Windrow (solar drying)	50,000/year
4. ARKANSAS			
1. Fayetteville (Northwest Arkansas Waste Mngt.)	Consideration (Co-composting)	A-SP	±400 cu.yd/wk.
5. CALIFORNIA			
1. Camarillo San. Dist.	Consideration	Vermicomposting	3
2. El Toro: Los Alisos	Operational	A-SP	.25
3. Fallbrook	Operational	Vermicomposting	1-1½
4. Hayward	Operational (by Miltona Brothers)	A-SP	10
5. Los Angeles (City)	Design (by San Joaquin Composting, Inc.)	Windrow	375 wet tons/day
6. Los Angeles (JWPCP County)	Operational	Windrow	
7. North County	Planning	In-Vessel (w/windrow curing)	25
8. Oakland: East Bay Munic. Util. Dist.	Operational	A-SP	60
9. Oxnard	Design (site selection)	In-Vessel (w/windrow curing)	30
10. San Diego	Pilot	Windrow	10 (50-100 at full-scale)
11. City of S. San Francisco	Operational	Windrow	10
12. Simi Valley	Pilot	Windrow	10 (full-scale)
6. COLORADO			
1. Denver Metro	Operational	Aerated Windrow	70
2. Ft. Collins	Operational	Aerated Windrow	6
3. Longmont	Design	A-SP	8.5
4. Summit County	Operational (by Twin Landfills)	A-SP	2
5. Wheatridge	Operational	Windrow	3/month

STATE PLANT NAME	STATUS	TYPE	SLUDGE VOLUME *dry ton/day* *(unless noted)*
7. CONNECTICUT			
1. Bristol	Operational	A-SP	10
2. Fairfield	Design	In-Vessel (IPS)	3
3. Farmington	Design	In-Vessel	20 yds./day
4. Glastonbury	Consideration	In-Vessel (IPS)	±1.5
5. Greenwich	Operational	A-SP	2500 cu. yd./yr.
6. Groton	Consideration (Regional facility—25,000 tons/yr.)	In-Vessel (IPS)	
7. Hartford	Construction	In-Vessel (American Bio Tech)	53-78
8. Killingly	Consideration (Regional facility)	In-Vessel (IPS or Royer)	7,000 wet tons/yr.
9. New London	Consideration (Regional facility)	In-Vessel (IPS)	5-6
10. Norwich	Pilot	Windrow	1-2
11. Ridgefield	Operational	Windrow	160,000 lbs/wk.
12. Windham	Consideration	In-Vessel (IPS)	1.7
8. DELAWARE			
1. Middletown/Odessa	Operational	In-Vessel (Laidig)	.5
2. Seaford	Operational	A-SP	.7
3. Wilmington (Del. Reclamation Project)	Operational	In-Vessel (Fairfield)	70
9. FLORIDA			
1. Brooksville	Design	In-Vessel (Bedminster Bioconversion)	
2. Collier County	Consideration	Windrow	
3. Cooper City Utilities	Operational	Windrow	(3.5 mgd)
4. Daytona Beach	Consideration		
5. Fernandina Beach	Operational	Windrow (Co-composting w/leaves)	
6. Fort Lauderdale	Operational	In-Vessel (Purac)	8
7. Gadsden Cty.: Florida State Hosp.	Operational	Windrow (Co-composting w/leaves)	
8. Hillsborough Cty. Utilities	Design	In-Vessel	35
9. Jacksonville: Buckman Plant	Operational	A-SP (Compost for odor filter)	2.26
10. Lee County	Consideration		16
11. Loxahatchee River Dist.	Operational	A-SP	
12. Mandarin County Utilities	Operational	Windrow	3
13. Margate	Planning	Windrow	
14. Meadowood Utility	Operational	Windrow	4/month
15. Melbourne	Planning	In-Vessel (Favored)	30 wet tons/day
16. Orange County (at landfill)	Operational	Windrow	
17. Orlando	Consideration		40-50
18. Palm Beach County	Planning	In-Vessel	±30
19. Plant City	Operational	A-SP	6
20. Reedy Creek	Operational	In-Vessel (Taulman)	9
21. Royal Palm Beach	Design (complete)	Windrow	12 wet tons/day
22. Sarasota	Operational	In-Vessel (Purac)	6-10
23. Univ. of Florida	Operational	Windrow	
10. GEORGIA			
1. Northeast Clayton County WPCP	Operational	In-Vessel (Taulman)	1.6
11. HAWAII			
1. Waimanalo WWTP	Consideration	A-SP	17
12. IDAHO			
None			
13. ILLINOIS			
1. DuPage County Public Works	Operational	A-SP	6-8
14. INDIANA			
1. Bloomington: Blucher Poole WWTF	Operational	Windrow	2
2. Crawfordsville	Operational	Windrow	18/month
15. IOWA			
1. Davenport	Pilot	Windrow	30 (full-scale)
2. Des Moines	Consideration (Co-composting)		40,000 wet tons yr.
3. Ft. Madison (Lee Cty.)	Operational	A-SP	15/week

SLUDGE COMPOSTING FACILITIES IN THE UNITED STATES—CONTINUED

STATE PLANT NAME	STATUS	TYPE	SLUDGE VOLUME *dry ton/day (unless noted)*
16. KANSAS			
1. Mission	Consideration	Windrow & In-Vessel	20
2. Topeka: Oakland WWTP	Operational	Windrow	10-12
2. Wichita: WPCP #1 & #2	Operational	Windrow	10
17. KENTUCKY			
1. West Hickam Creek WWTP	Operational	A-SP	4.5-5
18. LOUISIANA			
None			
19. MAINE			
1. Bangor	Operational	A-SP	1964/yr.
2. Bar Harbor	Operational	A-SP	365/yr.
3. Gardiner	Operational	A-SP	1
4. Kennebunkport	Operational	A-SP	50/yr.
5. Kittery	Operational	A-SP	35/yr.
6. Lisbon	Pilot (in 1989)		1
7. Old Orchard Beach & Saco	Operational	A-SP	3234/yr.
8. Old Town	Operational	A-SP	24/wk.
9. Portland	Operational	A-SP	Small quantity 16.2
	Design (Complete)	In-Vessel (IPS)	
10. Scarborough	Operational	A-SP	.75
11. South Portland	Pilot	Windrow	
12. Yarmouth	Operational	A-SP	219/yr.
20. MARYLAND			
1. Aberdeen	Design	A-SP	1.6
2. Baltimore: Back River	Operational	In-Vessel (Paygro)	48.3
3. Cambridge	Operational	A-SP	300 cu. yr/month
4. Elkton	Operational	A-SP	2
5. Havre de Grace	Operational	Aerated Windrow	2-3
6. Montgomery County	Operational	A-SP	40
7. Perryville	Operational	A-SP	2.5
21. MASSACHUSETTS			
1. Amherst	Planning	In-Vessel	8
2. Barre	Construction	A-SP	5 yds./wk.
3. Berkshire County (Northern)	Consideration (Co-composting)		
4. Berkshire County (Southern)	Consideration (Co-composting)		59 wet tons/day (sludge & septage)
5. Billerica	Design (complete)	A-SP	5
6. Boston	Consideration	All options	120-125
7. Bridgewater	Construction	A-SP	3.5
8. Dartmouth	Design	In-Vessel	7,300 lbs./day
9. Deer Island (MWRA	Pilot	A-SP	3
Edgartown	Planning (& Pilot)	A-SP	1
11. Gloucester	Planning (by Agrisource, Inc.)	A-SP	800/yr.
12. Haverhill	Design	In-Vessel	12-15
13. Hudson	Consideration	A-SP	60 wet tons/wk.
14. Leicester	Operational	A-SP	.20
15. Leominster	Design (complete)	A-SP	6
16. Mansfield	Operational	A-SP	143 yds/wk.
17. Marlboro	Construction	A-SP	12
18. Medfield	Design	A-SP	3
19. Nantucket	Design	A-SP	3
20. Orleans	Construction	A-SP	3.9
21. Pepperell	Construction	A-SP	8 yds./wk.
22. Rockport	Planning (w/Gloucester)	A-SP	1
23. Somerset	Construction	A-SP	2.5
24. Southbridge	Operational	A-SP	18 cu. yd./day
25. Springfield	Planning	In-Vessel (Taulman)	8,800/yr.
26. Swampscott	Operational	A-SP	1.4/wk.
27. Westborough/Shrewsbury	Operational	A-SP	3
28. Williamstown/Hoosac	Operational	A-SP	120 cu. yd./day
29. Yarmouth	Design (w/septage)	In-Vessel (agitated bin)	3

STATE PLANT NAME	STATUS	TYPE	SLUDGE VOLUME *dry ton/day (unless noted)*
22. MICHIGAN			
1. Battle Creek	Consideration	In-Vessel (Taulman)	25
2. Mackinac Island	Operational	Windrow	
3. Trenton	Consideration	All options (inc. agitated bin)	2
23. MINNESOTA			
1. St. Paul (Southwest Area Resid. Mngt.)	Consideration	In-Vessel	40
24. MISSISSIPPI			
None			
25. MISSOURI			
1. Joplin Landfill	Pilot (Co-composting)		
26. MONTANA			
1. Missoula	Operational	A-SP (EKO-Compost)	5+
27. NEBRASKA			
1. Beatrice	Operational	Windrow	1.5
2. Grand Island	Operational	Windrow	7-10
3. Holdrege	Pilot	Windrow	.5
4. Kearney	Operational	Windrow	7/wk.
5. Omaha: Papillion Creek Plant	Operational	Windrow	1,200/yr.
28. NEVADA			
1. Clark County San. District	Design	Aerated Windrow	42
29. NEW HAMPSHIRE			
1. Bristol	Planning	A-SP	
2. Claremont	Operational	A-SP	2.2
3. Dover	Design	A-SP	5
4. Durham	Operational	A-SP	1.97
5. Hampton	Consideration	In-Vessel	600/yr.
6. Lebanon	Operational	A-SP	1.5
7. Merrimack	Operational (In-Vessel/Co-composting under consideration)	A-SP	21
8. Milford	Operational	A-SP	.75
9. Plymouth	Operational	A-SP (In-Vessel:IPS in design—2 dt/day)	.25
10. Sunapee	Operational	A-SP	
30. NEW JERSEY			
1. Buena Borough MUA	Operational	A-SP	3.5
2. Burlington County	Design (Co-composting)	In-Vessel (Horizontal bin)	30
3. Camden County MUA: Delaware No. 1 WPCF	Design	In-Vessel	35-40
4. Cape May County MUA	Operational	In-Vessel (Purac)	20
5. Manville Boro	Construction (complete)	A-SP	1.5
6. Middletown Township	Operational	A-SP	2
7. Musconetcong	Design	In-Vessel	3.5
8. Pennsville	Operational	Windrow	1
9. Rockaway Valley	Consideration	In-Vessel	5.5
10. Sussex County MUA	Operational	A-SP (Totally enclosed)	12
11. Warren County MUA (Pequest River)	Construction	A-SP	.46
31. NEW MEXICO			
1. Albuquerque	Design	Aerated Windrow	15
32. NEW YORK			
1. Binghamton/Johnson	Construction	In-Vessel (Taulman)	10
2. Clinton County (Plattsburgh)	Operational	In-Vessel (Fairfield)	23-24
3. Endicott	Operational	In-Vessel (Taulman)	2
4. Guilderland	Operational	A-SP	1
5. Highland	Pilot	In-Vessel (Ag-Bag)	1
6. LeRoy	Consideration (Regional facility w/10 communities)	A-SP or In-Vessel	6
7. Lewiston	Consideration	In-Vessel	40 wet tons/wk.

SLUDGE COMPOSTING FACILITIES IN THE UNITED STATES—CONTINUED

STATE PLANT NAME	STATUS	TYPE	SLUDGE VOLUME *dry ton/day (unless noted)*
8. Lloyd	Pilot	A-SP	
9. Lockport	Consideration		40+ wet tons/wk.
10. Monroe County	Consideration (Co-composting)	In-Vessel	100
11. Niagara Falls	Consideration		17
12. Rennselaer County (Eastern)	Consideration (Co-composting)		1 wet ton/day
13. Rockland County Sewer Dist. #1	Consideration	In-Vessel	4.5-5
14. Schenectady	Operational	In-Vessel (American Bio Tech)	15
15. Tomkins County	Pilot	A-SP	4 cu. yd./wk.
16. Tri-Municipal (Woppinger Falls & Poughkeepsie)	Operational	A-SP	2
17. Yorktown Heights	Operational	Windrow	1
33. NORTH CAROLINA			
1. Charlotte	Consideration	In-vessel	30
2. Hickory, Newton, Conover & Catawba Counties	Construction	In-Vessel (Ashbrook-Simon-Hartley)	20
3. Morganton (Catawba River Plant)	Operational	A-SP	4.5
4. Raleigh	Planning	In-Vessel	20
5. Valdese	Operational	A-SP	1
34. NORTH DAKOTA			
None			
35. OHIO			
1. Akron	Operational	In-Vessel (Paygro)	73
2. Columbus	Operational	A-SP w/In-Vessel drying (Paygro)	18-30
3. Hamilton WWTP	Operational	In-Vessel (Ashbrook-Simon-Hartley)	17
4. Lake County	Operational	Aerated Windrow	7
36. OKLAHOMA			
1. Tulsa	Pilot	A-SP	
37. OREGON			
1. Newberg	Operational	In-Vessel (Ashbrook-Simon-Hartley)	3.5
2. Portland	Operational	In-Vessel (Taulman)	30
38. PENNSYLVANIA			
1. Evans City (RBW Industries—Regional project)	Design	A-SP	50 wet tons/day
2. Lancaster	Operational	In-Vessel (Taulman)	32
3. Philadelphia	Operational	A-SP	300
4. Scranton	Operational	A-SP	18.2
5. Springettsbury	Operational	A-SP	4
39. RHODE ISLAND			
1. Bristol	Planning		3
2. Jamestown	Operational	Windrow	350 cu. yd/yr.
3. South Kingston	Planning	In-Vessel	500 wet tons/mnt
4. West Warwick	Operational	A-SP	3-5
40. SOUTH CAROLINA			
1. East Richland County PSD: Gills Creek	Operational	In-Vessel (Taulman)	4-5
2. Myrtle Beach	Operational	A-SP	430/yr.
41. SOUTH DAKOTA			
None			
42. TENNESSEE			
1. Bristol	Bid Stage	In-Vessel	14.3
2. Nashville: Central Treatment Plant	Operational	A-SP	40
3. Nashville: Dry Creek	Operational	A-SP	5-6
43. TEXAS			
1. Austin	Operational	Windrow	10
2. Dallas	Pilot planned	In-Vessel	7-10 (30 at full-scale)
3. Fredricksburg	Operational	Windrow	
4. San Antonio	Operational (1/89)	Windrow	900 cu. yd./day
5. Trinity River Authority (Grand Prairie)	Design	Aerated Windrow	10

STATE PLANT NAME	STATUS	TYPE	SLUDGE VOLUME *dry ton/day (unless noted)*
44. UTAH			
None			
45. VERMONT			
1. Bennington	Bid Stage	In-Vessel	2
2. Central Vermont Region (inc. Barre & Montpelier)	Consideration		10
3. Chittenden County Region (Burlington area)	Consideration		
4. Connecticut River Valley Region (inc. White River Junction & Brattleboro)	Consideration (Two towns may consider interim static pile composting)		15
5. Rutland Region	Consideration		15
46. VIRGINIA			
1. Hampton Roads San. District	Operational	A-SP	65
2. Henrico County	Construction	In-Vessel (Purac)	18
3. Moores Creek WWTP (Charlottesville)	Operational	A-SP	45-50/wk.
4. Upper Occoquan	Operational	Aerated Windrow	7.5
47. WASHINGTON			
1. Monroe	Operational	Windrow	
2. Seattle METRO	Operational	A-SP	10
3. Southwest Suburban Sewer District (Miller Creek & Salmon Creek)	Operational (by Northwest Cascade)	A-SP	3
48. WEST VIRGINIA			
None			
49. WISCONSIN			
1. Portage	Operational (Co-composting)	In-Vessel	2.4
50. WYOMING			
None			
51. DISTRICT OF COLUMBIA			
1. Blue Plains	Operational	A-SP	40
52. PUERTO RICO			
1. Arecibo	Construction	A-SP	15
2. Mayaguez	Construction	A-SP	74

(This survey was conducted in October, 1988.)

International Sludge Composting

Diminishing landfill space, an urgent need for organic material in the soil, and continuing opposition to incineration are among the factors fueling sludge composting in other countries. The following is a brief overview of the level of activity:

North and South Europe: Representatives of Agro Developpement SA, a French sludge management firm, and Anjou Recherche visited operating European sludge composting plants during autumn 1987 for the Commission of the European Communities. The following is a brief description of these facilities, provided by Jean-Luc Martel of Agro Developpement, located in St. Quentin Yvelines:

Senigallia, Italy: This is a 300 m^3 B.A.V. in-vessel facility composting raw sludge on a seasonal basis. Sawdust (purchased for \$8/m^3) is used as a bulking agent; 1,500 to 2,000 m^3/year of 30 percent organic matter is produced and marketed by the municipality for \$20/m^3 to vegetable growers.

Pesaro, Italy: Raw sludge—350 m^3—at 20 percent dry solids is composted in a B.A.V. system on a seasonal basis. Sawdust obtained for \$4/m^3 is used as a bulking agent; 2,500 m^3/year of 30 percent organic matter is used for municipal purposes. The in-vessel option was chosen for Pesaro and Senigallia primarily because of tourism considerations.

Schio, Italy: A windrow system is used to compost digested sludge, with the organic fraction of screened refuse used as a bulking agent. Schio produces a high quality compost but has difficulty in distributing it because the plant is 100 kilometers from the nearest markets.

Valdemingomez, Spain: Digested sludge—30 dry matter tons/day—is mixed with screened pellets of dried sludge and windrowed from May to September. Private contractors operate the facility, producing 10,000 tons/

year. The material is sold to growers for $25/ton.

West Germany: Sludge composting is becoming more important in this country because municipalities are facing strict regulations and lack of landfill space. Three facilities were visited: Hilchenbach, which is using the B.A.V. in-vessel system, Horn Bad Meinberg, which is using the Gebruder Weiss/Kneer in-vessel system, and Lemgo, using a static pile process, mixing the sludge with organic material from shredded screened refuse.

Canterbury, United Kingdom: A 10 dmt/day static pile composting facility (utilizing the Rutgers strategy) is in its first year of production. Straw is used as a bulking agent. The Water Research Center in Medmenham is working on a small-scale co-composting sludge/refuse project. Report researchers of Anglian Water and Hensby Composts of Huntingdon, Cambridgeshire: "The process of producing compost from cereal straw and untreated liquid sludge has been developed from the method which has been found to be so successful in the production of mushroom compost. It consists of using cereal straw both to absorb the water component of sludge and to provide a structural framework for the compost, thereby allowing efficient aeration to occur."

Halweg, Netherlands: The static pile facility, operated under cover, composts 90,000 dmt of urban and industrial sludge with wood chips and wood bark.

France: There are six sludge composting plants in operation (there are other co-composting units and industrial projects). Together, they produce 12,000 to 15,000 tons of mature compost a year. At the beginning of the 1990's, a total production of 50,000 tons per year is expected from 12 sites.

Japan: The Japan Sewage Works Agency reports that as of April, 1988, there are 24 local municipalities that have sewage sludge composting facilities. Only 350,000 cubic meters of the country's annual sludge production of 2,200,000 cubic meters are currently utilized. Dr. Taigo Matsui, director of the agency, reports that sludge composting has been used more widely in recent years because interest has been growing in the use of organic fertilizer to recoup the fertility of soil which has been seriously strained and wasted by chemical-intensive farming. He adds that the popularization of composting technology has made it possible to supply high quality sludge-turned-compost in large quantity.

Of the 24 facilities, 19 are operated by the municipalities themselves. In the other cases, the municipalities have commissioned fertilizer manufacturers to produce compost from dewatered sludge they supply. The composting capacity of the municipally-operated plants ranges from 0.72 to 206 wet tons/day of dewatered sludge cake, with most plants handling three to 15 tons/day. Of the 19 composting plants, 10 are classed as the vertical type (in which feedstock runs vertically in the composting process), and nine are the horizontal type (in which feedstock travels horizontally). At eight composting plants, organic additives such as rice chaff were mixed into sludge, but the remaining 11 plants operated without such additives. Nine plants use slaked lime and ferric chloride as a flocculant in producing dewatered sludge cake; another nine plants use polymers, and the remaining one uses both.

Australia: At the Castlereagh treatment plant, west of Sydney, municipal operators and representatives from a private fertilizer company, AMG Industries Pty. Ltd, have combined in large-scale production and marketing trials aimed at extensive commercial distribution of sludge composted with sawdust wastes.

The treatment involves composting one volume of dewatered sludge from the St. Marys plant with three volumes of sawdust and other compost ingredients. Composting heats the material to approximately 60 to 65°C for 16 weeks. The compost heaps are turned weekly for aeration and thorough mixing. This process far exceeds the US Environmental Protection Agency standards and guidelines from which it was developed. About 300 tons of sludge compost from the Castlereagh trials are sold to local landscapers each month.

3

COMPOSTING ACTIVITY: SOLID WASTE

Nora Goldstein

MORE communities and counties are analyzing—and selecting—solid waste composting systems. The driving forces include:

1) Both environmental and economic concerns about mass burn facilities, as well as effective public and political opposition. For example, a 2,250 ton per day mass burn plant scheduled for San Diego was cancelled when a November 1987 referendum banned facilities burning over 500 tons per day of solid waste within three miles of hospitals and institutions such as schools. As a result, San Diego issued an RFP for non-burn alternatives for solid waste disposal, including composting.

In addition, more strict emissions standards for incineration, which are likely to emerge to protect the environment and appease citizen opposition, will significantly increase capital and operating costs and the amount of ash generated, making this option even less attractive.

2) The poor matchup of incineration technologies to the needs of rural communities that do not have sufficient waste flow to make a facility economical. In addition, energy markets are often nonexistent. At the same time, rural areas will be hard hit by new Subtitle D regulations for landfills, as many of these operations will not be in compliance and won't be able to afford the measures required to meet the new standards.

3) A more organized and substantive push by the solid waste composting industry to educate and promote this alternative. The process will soon be aided by having several operating composting facilities to demonstrate the viability of the technology and the marketability of the compost. State and local officials, especially in the Midwest, increasingly are attracted by the potential of solid waste composting and are funding research, pilot projects, conferences.

4) A realization by urban waste officials that a combination of recycling, refuse derived fuel (RDF) production, and composting has the potential to handle large volumes of solid waste, reduce landfill dependency, and perhaps ease environmental and economic concerns.

Key Issues

A 1988 nationwide survey of solid waste managers identified at least 40 communities or counties in the consideration, design, construction or operational phases of a solid waste composting project. At the end of 1988, there are six operating MSW composting facilities. Two—one in New Castle County, Delaware and the other in Portage, Wisconsin—are composting MSW with sewage sludge. The others—Fillmore County and St. Cloud, Minnesota, Sumter County, Florida and Dodge County, Kansas—are processing only solid waste. Table 1 provides the status of all projects.

In contrast to sludge composting, which really got its start in the 1970s, MSW composting has a longer history. In the 1950s and 1960s, a number of vendors were marketing systems and a handful of projects were started.

Overestimation of markets and very low tipping fees at garbage dumps contributed to the demise of those facilities.

The economic climate for MSW composting has substantially changed since those days with the escalation of landfill tipping fees. The second major difference reflects the growing concern about environmental safety and difficulty of siting burn facilities.

Several state governments have taken an increasingly active role in the development of solid waste composting. Massachusetts, for example, established a $7 million grant program to help communities conduct feasibility studies or construct projects for composting (MSW as well as other fractions such as leaves and wood waste). Minnesota has a capital assistance grant program where it provides 50 percent of the financing for MSW composting projects. Iowa, as part of its Groundwater Protection Act, also will provide capital assistance to composting projects. Its regulations provide a good base for the development of composting overall.

"Communities have to prepare comprehensive waste management plans that include an evaluation of alternatives," says Stu Schmitz of the Iowa Department of Natural Resources. "By 1994, they have to begin implementing those plans, and by 1997, the state will not permit landfills unless they are part of an alternative system. Generally speaking, there is more of a push throughout the state toward using composting and waste reduction as the alternatives because given the state's population, it is harder to do waste-to-energy projects."

Other states, such as New York and Florida, which in 1988 passed comprehensive solid waste management legislation, have set priorities that put reduction and recycling/composting at the top of the list, followed by incineration and landfilling. Other states have been passing bans on the disposal of leaves and yard wastes in landfills, and encouraging composting of these materials. Creating this "composting mindset" should have a positive impact on the development of MSW composting projects.

The difficulty with siting incinerators, combined with rising costs and uncertainty about air emissions and ash disposal, have caused many larger municipalities to put plans for mass burners on hold. "The anti-incineration sentiment is certainly helping the climate for MSW composting in Massachusetts," says Sumner Martinson of the state's Solid Waste Division in the Department of Environmental Quality and Engineering. "And it is very difficult to site new landfills. Almost every major municipality in the state is considering MSW composting as an option."

Considerations for System Design

The development of MSW composting in the United States is in a peculiar phase: The level of interest is outpacing the level of knowledge and operational experience. Solid waste planners, public works officials, consultants and even vendors agree that a primary drawback to designing an MSW composting facility is the lack of ongoing projects in the U.S. Tours frequently are arranged to visit installations in Europe and Japan. Differences in waste stream composition, however, make it difficult to draw direct comparisons.

Despite the lack of examples, municipalities and counties are proceeding with projects. Because there is an increasing number of vendors bringing new systems onto the market, waste managers need a mechanism to evaluate their various options. The first and most difficult step, says Joel Alpert of E&A Environmental Consultants, is for the community to define what it is they want in a composting system. "Is it reducing the amount of waste going to the landfill?" he asks. "Is it reducing costs? If a community can come to grips ahead of time with what they want to achieve and what is most important, then it becomes easier to evaluate vendors."

For example, if a community has run out of landfill space and is hauling waste to a disposal facility at great cost, it may want to minimize residuals. On the other hand, a community that has 10 years of landfill life left and wants to extend it to 20 years by reducing the amount of waste that has to be disposed may not be as concerned about residuals. The critical factor is knowing what amount of residuals is acceptable because that affects the type of system selected, the degree of recycling and upfront processing, and end markets for the finished compost.

Another consideration, says Alpert, is if a community wants to own the facility or work out a privatized package. "The vendors are breaking out into two groups," he says, "those that want to own and those that just want to sell equipment."

A number of municipalities and counties are in the process of conducting feasibility studies and preparing to issue RFPs. Luis Diaz of Cal Recovery Systems, Inc. in Richmond, California, outlines some key components of an RFP:

1. The vendor should have a functioning process facility similar to the one proposed both in design and capacity;

2. The vendor should have the financial strength to fulfill their commitment;

3. Performance criteria must be defined to specify what the system needs to do in terms of capacity and the amount of compost being produced, e.g. specify throughput, volume of finished compost.

4. Compost quality should be defined generally. Put in limitations to make certain the product will be of a certain quality, e.g. outline an acceptable chemical analysis and specify the level of impurities in the compost; and;

5. In the financial evaluation of the operation, don't assign compost a value until it is established in the market. Don't expect to have a tremendous return on the sale of compost.

Diaz adds that the issue of compost quality is difficult to determine, in part because of a lack of clear definitions from regulatory agencies about what is a usable compost product. Including some specifications, e.g. amount of compost generated and quality, will make the evaluation process easier, "enabling the municipality to compare apples to apples instead of apples to oranges," says Diaz. "Otherwise you will get two vendors who both say, for example, that they can process 400 tons per day of MSW. But one may have only 20 percent come out as compost while the other may do more separation and produce more compost—at a higher cost."

Other consultants believe that specifying the quality of compost that the vendor must meet is difficult, because of the unpredictability of the waste stream. "A vendor could ask, 'How can we guarantee the quality of the compost if you can't guarantee what's coming into the facility?' " says Joel Alpert. "On the chemical side, quality of the the compost will be affected by what you put in. On the physical side, you can add mechanical separation at the end to polish up the material."

Waste characterization studies, conducted in the predesign phase of a composting facility, give municipalities a better sense of the waste stream and potential contaminants. Composition studies need to take into account future bans and waste reduction laws that may be enacted. Breaking out the waste stream into commercial/industrial and residential flows also helps with facility design. Rice County, Minnesota, for example, as part of its feasibility study, visited every industry in the county to get a percent of glass, metals, cardboard and possible contaminants that would be coming into the recycling/composting facility. Questionnaires were mailed to the commercial sector to get a similar breakout of waste composition. This information will help in the design of the compost processing lines.

There are many other factors that plug into the design of an MSW composting system. They include: 1) recycling/source separation programs; 2) the degree of upfront processing; 3) chemical and physical contaminants; and 4) optimum biological conditions for composting. These factors are interrelated, and each has an effect on the end uses/marketability of the finished compost.

Recycling/Source Separation: There is no question that recycling is highly compatible with composting. Glass and plastic are common physical contaminants in MSW compost. By separating these materials—as well as ferrous, non-ferrous and newspaper—out of the wastestream in a recycling program there is less of a need for an intensive separation phase at the composting facility, fewer residuals after the composting process, and in general, generation of a cleaner compost product.

"You need a very good program of recycling, then composting becomes one of the recycling steps," says Clarence Golueke, senior editor of *BioCycle* and director of research and development for Cal Recovery Systems. "That way, you can process the solid waste stream with a non-thermal system." The costs of a recycling program, however, should not be figured into the cost/ton for composting. "For example, if it costs $50/ton for recycling—a process that helps to prepare the material for composting—that should be a separate cost."

The more that can be kept out of the waste stream via recycling, the more efficient the composting operation will be. "Keeping recyclables out of the system is better than having to hand pick them out," says Ron Albrecht of Ron Albrecht Associates in Annapolis, Maryland. Yet, several of the operating MSW composting facilities, e.g. Fillmore County and Sumter County use hand-picking in the front-end of the system because the rural population base makes curbside source separation programs difficult to implement. If hand-picking is part of the system, the belt speed must be configured to human capabilities. "Our time and motion studies for what people can hand-pick efficiently is a belt speed of 30 feet/minute," says Albrecht. "The upfront hand-picking needs to be treated as a separate system, and not designed around how fast the actual composting or processing equipment can accept incoming waste."

There is a debate in the field of MSW composting about the ability to remove physical contaminants from the compost after they have been pulverized or shredded in the front end of the system. "Half the fight in MSW composting is what you do up front," says Ron Albrecht. "The back end—the composting portion—is fairly well established. We don't see any major breakthroughs in ways to speed up that process. But the front end is where systems are falling down. For example, trying to pull glass out after it has been pulverized is difficult. And if you put sludge in, it becomes even harder."

Degree of Upfront Processing: The various composting systems on the market function differently with regards to the amount of up-front processing, i.e. shredding, grinding, pulverizing. The optimum particle size for composting depends on the moisture content of the MSW. "The allowable minimum is about one inch," says Clarence Golueke. "If the particle size is too small, the composting mixture becomes fibrous. If the MSW is drier, however, the size can be smaller. But if one has a lot of wet vegetative material, you would end up with soup. With systems that include a bulking material, e.g. sawdust or straw, the permissible particle size may be one-quarter to a half-inch."

Physical/Chemical Contaminants: As was discussed earlier, the amount of physical contaminants in the compost will affect the amount of residuals and the marketability of the compost. Chemical contaminants can definitely affect end use, and could possibly hamper the biological process of composting. Establishing a household hazardous waste program to collect contaminants and keep them out of the waste stream going into the composting facility is an essential part of any composting operation. Efforts should be made to divert hazardous wastes from the industrial and commercial sectors as well. Burlington County, New Jersey, which is designing a co-composting facility, is developing an impressive multi-pronged approach to maximize the elimination of contaminants. (This and other projects will be discussed in the next issue.)

Optimum Biological Conditions: The art of MSW composting, unlike sewage sludge composting, is as much mechanical as biological. "The key to MSW composting is not the biological process, but separation," says Luis Diaz. "If you could keep all the organics from the inorganics, there would be no problem."

From a biological standpoint, most of the common physical contaminants that can get into the composting phase would not inhibit any biological reaction of the organisms, says Clarence Golueke. The exception would be concentrations of metals high enough to pose a problem. "As long as you don't create inhibiting conditions, there is nothing inherent to stop the biological process," he explains. "But the physical contaminants may make it difficult to apply the conditions needed for composting, e.g. proper temperatures, optimum aeration, and adequate moisture—especially if they hamper the workings of the composting equipment."

Unlike sludge or leaves and yard waste, MSW is a highly complex material, ranging from cabbage leaves to paper, that decomposes in stages. "There is a succession of bacteria," explains Golueke. "Each uses one kind of material as a substrate, which in turn serves as food for the next microorganism down the chain. If that succession gets interrupted by making some of the nutrients inaccessible, then total decomposition and stabilization will not occur."

Because MSW composting requires a combination of engineering and biological know-how, system designers and equipment suppliers should have people on staff who have training in these disciplines. "You need biological input by a biologist with an engineer to give the system a practical cast," says Golueke. "In the past, there have been damn few biologists among system designers."

Agitation during composting is recommended by some in the composting field to keep the material aerobic and to assist in the reduction of particle size. Overagitation, however, may provide too much aeration of the material. Again, knowledge of the biological process of composting will help to achieve optimum conditions.

Market Assessment

Markets for MSW composts have not been clearly established, primarily because of the lack of finished product with which to conduct experiments and generate data for end users. Much of the compost generated to date has been used for landfill cover. Operating facilities have the capability to produce a compost for other end uses, and some have been working on that aspect of market development.

Available markets are dictated by impurities in the compost, which in turn are dictated by the degree of upfront separation and back end finishing. "If the product is separated correctly and screened so you have the appropriate least amount of ferrous, non-ferrous, plastic and glass, and if the material is composted correctly, there are very important uses," says Roger Tuttle, president of Compost Management, Inc., a marketing firm in Doylestown, Pennsylvania that recently received a 20 year contract to market MSW compost from a facility to be built in Broward County, Florida. These markets include horticultural and landscaping applications, as well as blended soil amendment products.

Research comparing sludge, MSW, and leaf composts—conducted for Compost Management—found that the materials can not be used interchangeably. "The three composts were added to three types of soil—good, poor, and sandy," says Tuttle. "The water holding capacity, percent organics and cation exchange were completely different for each one, which means there are three different sets of parameters for product end use."

Municipalities and counties that have adequate time to get a full-scale composting project up and running may want to start first with a small-scale pilot. In addition to learning about the separation, processing and biological hurdles—and thus gaining knowledge about system design—market testing can be done with the finished compost.

Strategically speaking, adds Roger Tuttle, a community may want to use the MSW compost as landfill cover in the short term. Over the long term, markets can be developed and the compost can be upgraded. "That's how markets for sludge compost evolved," says Tuttle. "The original idea was to compost sludge as a mechanism of pathogen reduction and then landfill it. But researchers realized if the finished compost were screened, they would have a marketable end product."

George Fitzpatrick, a horticulturalist with the University of Florida's Research and Education Center in Ft. Lauderdale, has conducted several experiments with MSW compost. The first involved testing three composts—one from a facility in Dodge County, Kansas, another from a co-composting facility in Delaware, and the third produced from wastepaper sludge in Pennsylvania. The composts were used as a potting mix and compared with a control made from a commercially prepared soil-less potting mix. The three crops used in the experiment were the cuban royal palm, orange jessamine, and dwarf oleander. All three crops are grown commercially in tropical/subtropical climates and used mostly for landscaping.

"The results, in a nutshell, were that the growth rates with the palm and jessamine were not significantly different than the control," says Fitzpatrick. "But with the dwarf oleander, the three MSW composts performed better than the control. And the wastepaper compost did better than the other two. There appears to be a relationship between the soil pH and growth in the oleander, which is an alkaline sensitive plant." He concluded that the MSW composts were no better or worse than the commercial soil-less potting mix that he bought for $35/cubic yard.

Fitzpatrick's second experiment involved testing a sludge compost and an MSW compost as top dress in a landscaping job on the beach. The crop being grown was sea lavender. "We found that the MSW compost had two times the water holding capacity of sludge compost, which had 10 times the capacity of sand," he says. "And the grass had two times the survivorship in MSW compost than the sludge compost. The sand had no survivorship. The water holding capacity makes the MSW compost great for erosion prevention."

In short, Fitzpatrick believes that there is good, high value market potential for a screened MSW compost. "If it is unscreened, it can be used as landfill cover or artificial top soil," he says. "If it is screened, however, it can be used in the fertilizer blending market or in a potting soil as a growing medium."

He would like to focus future research in two areas: the influence of the feedstock on the end product and what influence the processing of the feedstock, e.g. grinding or shredding, before and after composting, has on the physical and chemical parameters of the end product. "These factors ultimately affect the salability and uses for the material," says Fitzpatrick. "Some facilities separate or plan to separate a lot, while others do very little. With regards to the effects of processing, I believe that the water holding capacity is closely related to pore space, which is affected by milling or not milling the final product. For growers, the particle size has to be in a certain range to optimize the water holding capacity."

Future Outlook

The level of interest in MSW composting, combined with the capability of the technology to process a significant portion of the waste stream and generate a quality end product, creates a very positive outlook for the future. There are many hurdles to jump, among them keeping chemical contaminants out of the compost feedstock and minimizing residuals that have to be landfilled. The critical factor at this stage of development is to get high caliber projects up and running, projects that include a realistic assessment of processing capacity and markets. States with grant programs can assist this process by funding pilot projects and helping communities evaluate their various options.

From a regulatory perspective, standards for finished composts must be established. Agricultural experiment stations, university horticultural departments, and others need to get involved with research on the effects of possible contaminants and potential uses for MSW composts. In addition, a number of counties and municipalities are evaluating or proceeding with a co-composting facility, processing MSW and sewage sludge together. State regulations need to reflect this trend and not eliminate possible markets for a quality finished product.

Managing solid waste today has boiled down to two essential goals: preserving landfill capacity and protecting public health and the environment. A combination of recycling and composting has great potential to take many waste managers in the direction of meeting those goals.

TABLE 1

Nationwide Listing of MSW Composting Facilities

Facility	Status	System Selected	Tons/Day (unprocessed MSW)
California			
1. Los Angeles	Consideration		
2. San Diego	RFP Issued		500 minimum
Delaware			
1. Wilmington (Del. Reclamation Project)	Operational	Fairfield	1,000 MSW (into RDF/Comp.) 70 dry tons/day of sludge
Florida			
1. Brooksville	Design	Eweson (Bedminster Bioconversion Co.)	40 (100 design) plus sludge
2. Broward Cty. (Pompano Beach, Dania, Hallandale, Pembrooke Pines)	Design	Buhler-Miag (Reuter)	600
3. Dade Cty.	Design	Agripost, Inc.	800
4. Sumter Cty.	Operational	Envtl. Recovery Systems (windrow composting)	100 (300 design)
Illinois			
1. Champaign Cty.	Consideration		
2. McHenry Cty.	Pilot planned	Windrow	10-15 (400 full-scale)
Iowa			
1. Cherokee Cty.	Construction	Lundell (w/silo for composting)	30
2. Council Bluffs	RFP issued		40,000/year
3. Des Moines	RFP issued		300
4. Harden Cty.	Design	Privatized using Lundell	

TABLE 1—*continued*

Nationwide Listing of MSW Composting Facilities

Facility	Status	System Selected	Tons/Day (unprocessed MSW)
Kansas			
1. Dodge County	Operational	Trash Reduction Systems (windrow composting)	30
Massachusetts			
1. Boston	Consideration		800
2. Berkshire Cty. (Northern)	Feasibility Study for co-composting		
3. Berkshire Cty. (Southern)	Feasibility study for co-composting		100 plus sludge
4. Hampshire Cty.	Feasibility study for co-composting		
5. Lowell	Consideration		
6. Nantucket	Vendor negotiation	Buhler-Miag (Resource Control)	100 MSW sludge
Minnesota			
1. Albert Lea	Pilot completed	Windrow	
2. Carver/Scott Counties	Feasibility Study		200
3. Crow Wing Cty.	Design	Dano (Waste Processing Corp.)	120
4. Eden Prairie	Permit Applic.	Buhler-Miag (w/RDF system)	400
5. Fillmore Cty.	Operational	Windrow	15-20 (30 design)
6. Hennapin Cty.	Consideration		200
7. Lake of the Woods Cty.	Pilot	Windrow	2500/year
8. Martin/Jackson/ Faribault Counties	Feasibility Study		80–100
9. Rice Cty.	Feasibility Study		38,000–45,000/year
10. St. Cloud	Operational	Dano (Recomp, Inc.)	60
11. Swift	Design	Windrow	15
12. Washington/Ramsey Counties	Consideration (w/RDF system)		
13. Wright Cty.	Vendor selection		100
New Jersey			
1. Burlington Cty.	Design	Horizontal bin-type system	200 (processed MSW) 150 wet tons of sludge
New York			
1. Easthampton	Pilot completed	Windrow	
2. Rennselaer Cty.	Feasibility study for co-composting		100 (both MSW & sludge)
3. Southold	RFP issued		120
North Carolina			
1. Buncombe Cty.	Feasibility study		100,000/year
Ohio			
1. Shelby Cty.	Permit applic.	WPF Corp.	200
Oregon			
1. Portland	Vendor negotiation	Dano (Reidel)	185,000/year
Texas			
1. Big Sandy	Pilot	Eweson	
Wisconsin			
1. Portage	Operational	Drum w/windrows	100/wk. of MSW 50 tons/wk. of sludge at 3% solids

International Solid Waste Composting

There has been more steady activity of MSW composting in other countries than in the United States. In fact, many state and local government officials in the U.S. travel to other countries to view operating MSW composting facilities as part of their own feasibility study and design process.

The following is a sampling of MSW composting activity:

France: There is a regional plant located outside St. Brieuc in the Chatelets region of France that receives 25,000 tons per year of urban solid waste generated by 20 townships. Prior to composting, the waste is mechanically preprocessed (e.g., using a hammermill, magnetic separator, trommel, and air classifier). The composting is done using the Siloda process, an accelerated technique. The facility contains four silos, each with a unit capacity of two days' production. The fermentation process lasts from eight to 10 days, after which the material is transferred to a curing area using a paddle wheel designed by OTV. The compost remains in the curing area for two months. A grinder and rotary screen are used for final treatment—to eliminate residual plastic scraps. The output of refined compost is about 10,000 tons per annum.

Greece: Both the city of Athens and the city of Heraklion (on the island of Crete) have been conducting pilot studies of MSW composting. The Athens project includes an RDF process, recovery of ferrous metals, and composting. Heraklio used the windrow composting method and extensively tested the application of the end product in greenhouse experiments.

Germany: The City of Duisburg has a 30-year-old composting plant that processes household waste from 95,000 residents. During the three autumn months, the plant also treats leaves from the entire city as well as manure from the slaughterhouse and zoo. Prior to composting in Dano drums, the refuse passes through a magnetic separator; glass bottles, non-ferrous metals, hard plastic and voluminous items are sorted out manually. Optimum moisture content in the drum is obtained by adding sewage sludge from a nearby plant. Following an average retention time of 36 hours in the drum, the material is screened. The compost portion is divided into two grades (leaf and refuse) and further matured on an aerated bed. Two primary markets for the compost are viticulture and landscaping.

A group of Michigan officials traveled to Europe recently to tour composting facilities in an effort to determine the feasibility of using compost technology to process MSW. Wayne Koser of the Michigan Department of Natural Resources, gave this overview of some of the operations visited in Germany:

- Wittzenhausen: This community of approximately 20,000 inhabitants was the first in Germany to adopt a source separation program for all organic wastes supplemented by a backyard composting program. A great deal of research is being conducted in testing various seepage collection systems, temperature retention times, turning cycles and mixing rates of green waste with various organic wastes.
- Bad Kreuznach: This facility services 14 districts and 119 assorted villages, processing over 220 MT/day of MSW. It utilizes a Dano system, and is the newest composting facility in Germany.
- Heidelberg: The central unit of the Multi-Bacto Compost/Incinerator Plant here consists of four rotting towers. One tower is devoted to composting 15 MT/day of source separated organics; the three other towers are used for composting 45 MT/day of mixed MSW comprised of 10 to 20 percent sewage sludge.
- Singen: The Brikollare Composting Facility is designed to handle 70,000 MT/year of household MSW and 20 MT/year of dehydrated sewage sludge. It serves a population of 250,000 inhabitants. The program includes household hazardous waste collection.

Switzerland and Belgium: The Michigan group also visited sites in these countries. In Schaffenhausen, Switzerland they saw an integrated solid waste management facility comprised of three component parts: a sewage sludge processing section, a Buhler-Miag compost unit and a waste-to energy incinerator. There is a small (8 MT/day) demonstration plant in Leibstad that processes mixed household/commercial waste. Two composting facilities were toured in Belgium. At an integrated facility in Tournai (designed like Schaffenhausen and Heidelberg), recyclables are removed by site separation, organics are removed for composting and rejects and unrecyclables are incinerated. In Ghent, the group observed the Dranco Methane/Composting Pilot project which is designed to utilize approximately 3 MT/day of organic waste in a dry anaerobic composting process that produces a product of methane- and compost-like substances.

One observation made by the group was that compost from programs that mixed MSW and sewage sludge contained the highest levels of heavy metals and other forms of contaminants, creating a major environmental concern as to their use. Those programs that involved source separated organics were the most successful. The composts had considerably lower levels of heavy metals and other contaminants and were much easier to market with few, if any, environmental restrictions as to their use. The group concluded, therefore, that if MSW composting is done in concert with a comprehensive integrated solid waste plant and a source separation collection program, it can serve as a viable option to current landfill practices in Michigan and the United States.

Ecuador: A composting facility was built in 1956 by Dano for La Fabrica Nacional de Abonos Ecuador, originally for use as a fertilizer factory. Eventually, the plant was taken over by the city, which now operates it 16 hours/day, selling all the compost produced. Ferrous metals are extracted from the waste stream prior to being fed into a Dano rotating drum. The moisture content ranges (by season) between 50 and 85 percent, so no water or sludge is added prior to composting. The material is screened after exiting the drum (moisture content is between 30 and 40 percent), and is taken to a maturation area. After being dried to 20 percent, mature compost is screened to minus 12 mm. The compost is then blended with guano or urea to enhance N content and sold for the equivalent of $58/ton.

New Zealand: The City of Auckland contracted with a private company to collect and compost its MSW. The company plans to use the Ebara system, constructing a 200 ton per 8-hour day facility on five hectares of land owned by the city.

Yard Waste Composting

The intensive push toward landfill avoidance in the U.S. and other countries is leading toward a boom in leaf and yard waste composting facilities. A *BioCycle* survey conducted in mid-1988 found literally hundreds of communities throughout the U.S. with operating projects. The boom has been fueled in some states by legislation that bans the disposal of leaves and other yard wastes in landfills. In New Jersey, for example, which passed a ban that covers only leaves, there were 180 permitted sites. A state official expected that number to rise to 300 by 1989.

As one would expect, the amount of yard waste available in any community varies widely from one part of the country to another. Most waste composition studies put the percentage at anywhere from 10 to 30 percent of the MSW stream. During the fall, however, leaves can represent 60 to 80 percent of the waste stream. During the summer, grass clippings also represent a significant portion. Omaha, Nebraska, for example, estimates that approximately 50 percent of the residential waste picked up by its crews during the summer is grass.

In addition to the significant diversion potential, yard waste utilization projects can be developed in a very economic manner. Communities that already separately collect yard waste simply have to redirect collection vehicles from a landfill or waste-to-energy facility to the utilization program. Most yard waste projects include a composting facility to process the material. Another approach, depending upon location, is to use the yard waste directly, e.g. applying leaves to nearby farmers' fields. For example, Lower Swatara Township in Pennsylvania provides its leaves to farmers, who use the material for animal bedding. Many communities used chipped woody material for mulch and park trails.

Another very decided advantage to yard waste utilization is the market situation. In recycling programs, most of the material collected is sold on the open market and thus must compete with materials from other programs—and ride the ups and downs of external markets. The majority of the current yard waste projects, on the other hand, operate in a closed loop system. The material generated is utilized within the community from where it came. In most cases, either the municipality uses it for landscaping purposes or cover at the local landfill, or it is given back to citizens for their use.

Success at individual community yard waste utilization projects has led a number of states to formulate strategies aimed at encouraging the development of even more programs to help alleviate disposal problems. The efforts have taken a number of forms. The most dramatic is prohibiting the disposal of yard waste in solid waste disposal facilities other than composting sites. Minnesota and Wisconsin recently passed legislation that would ban the disposal of yard waste, including leaves, grass clippings, garden debris and brush, from disposal facilities before 1993. New Jersey's ban, which took effect in April 1988, only covers leaves and not other types of yard waste.

A less drastic approach used to encourage yard waste composting is to allow those facilities to be permitted and operate under less stringent regulations than ones that apply to other types of solid waste processing facilities. Most states that take a less stringent approach do so because they view yard waste as a resource rather than a waste, in much the same manner that source separated paper, glass and metal are viewed as a resource. Many state regulatory agencies also believe that the potential for environmental problems at yard waste composting facilities is much lower than at mixed waste composting facilities and other waste processing systems.

Industrial And Hazardous Wastes

Papermills, breweries and food processors are among the industries composting their non-hazardous industrial wastes. Traditionally, these by-products have been landfilled, stockpiled on site, or in some cases, applied on land. Experimentation is also being done on the viability of composting some hazardous wastes, such as insecticides in agricultural wastes.

The State of Maine has been actively looking into ways to treat industrial wastes through composting. A list compiled by the state's Department of Agriculture is long and varied: papermill sludge (the highest in quantity produced in Maine), apple pomace, wood by-products, wood ash, potato processing wastes, kiln cement dust, animal manure, seafood wastes, and other more esoteric materials. Massachusetts has an active program for composting agricultural wastes, and Wisconsin is looking into alternatives to landfilling papermill sludges. Detailed information on some of these projects is included in a later chapter.

SECTION TWO

DESIGN AND OPERATIONS

4

LEARNING FROM EXPERIENCE

THERE are many challenges to making the operation of a composting facility a success. One key to meeting these challenges head-on is to learn from the experiences of operating facilities. "Once a municipality puts money into a system, it is hard to put more capital in to change it," says John M. Walker of the U.S. Environmental Protection Agency's Office of Municipal Pollution Control in Washington, D.C. "So you have to be careful and cautious the first time around with equipment investments."

Spending time learning about other operating facilities not only gives municipalities and companies who are designing a plant a good orientation, it also points out the important factors in a composting operation. The ideal situation, says Walker, is for selected members of the future operating staff to work at an existing plant for a short while. "The amount of money a municipality would spend for that is infinitesimally small compared to what it costs to try to operate and retrofit an inappropriately designed plant. It's politically difficult, however, to get approval to work at or inspect other facilities. But getting that experience is a small investment that pays great dividends."

After observing composting facilities for many years, Walker says that one essential component for success is flexibility. It starts with having a disposal backup in case there is a delivery of sludge from the wastewater treatment plant that should not be accepted because of an initial septic state, undesirable contaminants or temporary problems at the facility. Some considerations with regards to flexibility are:

1) Being able to deal with the inconsistencies in sludge solids and bulking agents. "The bulking agent itself is inconsistent; size and volatile solids content will vary," says Walker. "And the sludge solids content varies constantly, forcing a regular change in the ratio of sludge to bulking agent. Facilities are always designed for drier sludges and inevitably it comes in wetter. Whatever system you have, it must be able to handle that variation. And sometimes, when mixing the two, a blend may not be right. There should be a way to adjust it, for example, to get in there with a front-end loader and change the mix."

2) If mechanized, having other ways to keep the system going. An aerated static pile facility in South Portland, Maine was designed in such a way that the only way to feed the pug mill was by the conveyors from the sludge hopper and wood chip silo. When the conveyors didn't operate properly or the silo froze, there was no other way to load the pug mill.

"In general, the more mechanized a facility is, the more difficult it will be to provide for flexibility," says Walker. "Components of the system are usually at fixed locations and are difficult to change around. On the other hand, a facility that's mechanized can be much easier to operate."

3) Being sure to have a backup. "Whether you are running a composting facility or an incinerator, you need a backup—a landfill or a land application site," says

Walker. "Also, it's very important to have easy access to do repairs. Part of good management is being able to get in and do fairly rapid maintenance. If a system breaks down, the odors start up right away."

4) Adjusting to changes occurring at the wastewater treatment plant. Changes in sludge characteristics affect the actual composting process.

"In Los Angeles County, for example, the sludge character has changed a lot in the last five years," says Walker. "Wastewater from a paper processing plant was coming into the treatment plant with a high cellulose content winding up in the sludge. The paper company started pretreating its wastewater and the cellulose level decreased significantly. At the composting site, the operators began to notice that they were not getting the normal heating levels in the windrows necessary for pathogen destruction. The energy available for exothermic microbial activity decreased with the loss of cellulose. Hence, they compensated for the lack of the cellulose energy source by going from smaller to larger windrows to gain insulation and higher heating levels."

Improved anaerobic digestion prior to composting at the Los Angeles County facility further removed volatile materials that would have been a source for exothermic microbial activity.

Changes in the dewatering process at the treatment plant can affect both the composting process and the end use of the product. Dewatering generally yields sludge moisture contents that vary and that average up to 5 percent lower than anticipated during the design of the composting system. Full capacity operation is then difficult at best with more bulking agent, space, and money required to compost a given volume of sludge. There is also a greater potential for malodor.

Chemical additions to the sludge during dewatering also affect the composting process. For example, says Walker, treatment plants that use certain ferric chloride waste products as a dewatering aid or to help remove phosphate may be adding heavy metals into the sludge that can hinder the utilization of the compost. Addition of lime for dewatering also has an impact on the compost process by lowering the volatile solids content and influencing the end use options for the compost. While a high lime composted sludge product is good for use on many acid soils and causes reduced availability of heavy metals, it is not good for the growth of acid-loving plants. Composting facilities have to be aware of these changes, not only for sludge processing but also for marketing considerations.

Phased-In Start-up: Putting components planned for a new system through some form of pilot testing is essential prior to incorporation into the design. This testing offers a chance to review materials handling procedures and equipment functions. However, says Walker, be careful about applying the results of a pilot test to full-scale operation. "Other problems might arise when you load up an operation to full capacity. The mechanical systems are then fully stressed; the air supply, distribution and scrubbing components are strained; and difficulties with controlling odor often occur."

Some enclosed systems have had aeration difficulties when operated at full capacity. In the pilot stage, when the vessel was not loaded to the top, air could be blown through. But when loaded at full height there was not enough air capacity or it was too expensive to blow air through.

In general, the ideal situation is to pilot test, then operate at a lower capacity, and finally, operate the system at full capacity. This will help to ensure that all the components are operating properly.

Operational Factors

When designing a sludge composting facility, a variety of factors need to be taken into consideration. Among these are sludge quality, dewatering, bulking agents and amendments, process control, and odor control. This section provides an overview of some of these essential areas. More detailed discussions on process control are included in a later chapter.

Sludge Quality

Facilities are successfully composting both raw and digested sludges. In 1986, the Clark County Sanitation District in Las Vegas, Nevada conducted a small-scale windrow composting demonstration project utilizing raw organic sludges, for the purpose of answering questions concerning odor potential and product quality. In addition, the county believed that substantial capital cost savings could be realized if the construction of anaerobic digesters could be eliminated from the sludge management process train.

The induced aeration windrow composting process was utilized in the pilot study. Seven windrow cycles were completed during the study period in 1986. There were variations in bulking particles (primarily wood chips and sawdust were used), aeration rates, combinations of raw and digested sludge, and addition of conditioners, i.e. lime and ferric chloride. The following conclusions were reported by James Wren-Jarvis, the District's Deputy Director:

(1) Raw sludges conditioned with lime and ferric chloride prior to filter press dewatering can be successfully composted in aerated windrows.

(2) Raw sludges conditioned with polymers prior to dewatering tend to produce a more odorous filter cake, which can be successfully composted with carefully monitored and controlled aeration rates.

(3) Frequent turnings are necessary to bring the cooler outer layers of the windrow in contact with the higher internal temperatures.

(4) Raw dewatered sludges with 30 percent total solids content could not be composted without the addition of a bulking material to provide free air space, porosity and structural stability.

(5) Study windrows required peak aeration rates as high as 187m^3/metric dry ton of sludge to control peak demands for temperature control.

(6) Windrows with initial total solids concentrations of 35 percent, although slower in starting, were ultimately composted successfully.

(7) When temperatures of 55°C and above are maintained for 15 days with five turnings, adequate disinfection is provided to the compost material.

(8) Temperature elevations occur more rapidly in polymer conditioned sludges than lime stabilized sludges.

(9) Carbon dioxide and ammonia production in aerated windrow composting produced by slow microbial activity eventually neutralizes high or low pH materials in the windrows.

(10) Natural material beds for odor control may have an application in remote areas, but for urban locations the need for mechanical odor scrubbers seems apparent.

Dewatering

The total solids content of sludge is one of the most important factors affecting composting efficiency and economics. Sludge solids have an impact on the materials balance and therefore the facility size, labor force, materials handling equipment, and overall operating and maintenance costs, points out Dr. Eliot Epstein of E&A Environmental Consultants. Having a consistent solids content is critical to a successful composting operation, therefore facilities want to pay careful attention to the dewatering phase of the sludge management process.

Data from *BioCycle*'s annual compost facility survey shows that belt filter presses are used most frequently. Other methods of dewatering include open or covered sand drying beds, mechanical spreading and turning, lagooning of wet sludge, vacuum filtration, and centrifuges. Chemical usage in conditioning and dewatering can have an impact on the quality of the final compost product, and thus potential applications. Explains Dr. Epstein:

"Belt filter presses utilize polymers in contrast to the usage of lime and ferric chloride in vacuum filters or presses. Lime, ferric chloride, and alum will increase the content of soluble salts in the compost. Reduced plant growth from high soluble salt levels is well documented. Horticultural and greenhouse plants are particularly sensitive to salts and continuous monitoring is necessary in the production of these plants in order to prevent damage to plants. The use of polymers in the dewatering process does not affect the process or the product and their usage is preferred."

There is a great deal of innovation in the field of sludge dewatering. These developments include:

• Sludge Freezing: The Cold Regions Research & Engineering Laboratory in Hanover, New Hampshire has been conducting research into using sludge freezing in the winter for conditioning and dewatering. This approach, applied in cold climates, is combined with high rate sand bed systems in the warm months. Sherwood Reed of the U.S. CRREL explains the system:

Sand drying beds have been the traditional method for sludge dewatering in small to moderate sized communities. Under favorable conditions, they can still outperform, in terms of final sludge cake solids, the more modern and more complex mechanical dewatering methods. Because the weather does not always cooperate, sand beds in the northern half of the U.S. are either a seasonal operation and/or are covered by some type of greenhouse. In addition, alum and other chemical sludges are difficult to dewater on sand beds. The combination of these factors has led to the trend in recent years away from sand beds and toward more mechanical and energy intensive systems.

Freezing and then thawing a sludge will convert a material with a jelly-like consistency to a granular type material that drains readily. Solids concentrations exceeding 20 percent will be realized as soon as the material is thawed, and 50 to 70 percent can be achieved with minimal additional drying time. The process will work with any type of sludge, at any solids concentration, but is particularly effective with chemical and biochemical sludges which do not drain readily. Energy costs for artificial freeze-thawing are prohibitive so the concept must depend on natural freezing to be cost effective. The feasible area can include most of the northern half of the United States with the exception of the West Coast.

It is essential for the layer to freeze completely if the dewatering benefits are to be realized. In many locations a large single layer may never freeze completely to the bottom, with only the upper portion going through alternating freezing and thawing cycles.

• Reed System: The Max Planck Institute in West Germany developed a dewatering process using the common reed *Phragmites*. It combines the action of conventional drying beds with effects of aquatic plants on water-bearing substrates. Several towns in New Jersey are using the reed system. The plants are reported to desiccate the sludge deposits and also cause extensive mineralization. Tests at a West German treatment plant demonstrated the ability of the root system to dry and compact sludge by continuously absorbing water from all strata of the sludge. Vertical stems of the plants provide channels for continual drainage of water; mineralization is caused by a microflora that establishes itself in the root system of reeds.

• Electroacoustic Dewatering: Researchers at Battelle Memorial Institute in Columbus, Ohio have been working on an electroacoustic dewatering process to augment conventional devices such as vacuum filters. The process, says Dr. Satya Chauhan of Battelle, enhances water removal by novel solid/liquid separation techniques that greatly reduce the energy needed for drying and thus lower overall sludge processing costs.

In Battelle's process, when an electrical current is applied, charged particles are directed away from filter surfaces reducing blinding of the filter, and water moves in the opposite direction under electroosmotic forces. At the same time, acoustic energy sends waves through the material, creating forces that promote diffusion and migration of the liquid from the particles.

Dr. Chauhan points out that the patented process has much application in composting facilities, since solids content can be increased to 35 percent, thereby reducing the amount of bulking agent required. He believes that the process is useful for medium to large municipal wastewater treatment plants.

Explains Dr. Chauhan, manager of Battelle's Advanced

Separations Program Office: "The electroacoustic dewatering process can produce solids contents in sludge exceeding those required for sustaining combustion without use of supplemental fuel. Also, sufficiently dry sludge can be produced to eliminate the need for blending sludge with wood chips for composting applications. We have successfully scaled up the process to about 15 lb/hr scale as far as the municipal sludge dewatering is concerned."

• Solar Drying: A Phoenix, Arizona wastewater treatment facility tested solar sludge drying techniques for several years. Digested sludge is pumped to soil cement-lined, shallow drying lagoons equipped with gravity decant facilities for supernatant and storm water removal. Sludge drying to 75 percent solids content is accelerated by daily mixing with a turner-composter to maintain continued exposure of free moisture to the atmosphere. When sludge is dewatered sufficiently to form windrows, the mixing operation conforms to standard windrow composting techniques and the elevated temperature of the aerobic composting process aids in sludge drying.

Another approach to solar sludge drying has been developed by staff at the Sandia National Laboratories in New Mexico. The solar dryer includes a "greenhouse," a crossbar conveyor and 12-inch ports for aeration. Sludge at 8 percent dry solids enters the greenhouse and moves across the elevated drying bed (7½ × 48½ feet). Deeper sludge beds can be leveled and mixed with eight paddle bars suspended three inches above the drying bed.

Water vapor is removed from the module by 12-inch ports which connect the interior with an exterior condensation system. Water is extracted through an underfloor drain network.

• Mobile Systems: Because of rapid growth, wastewater treatment plants often need to develop additional sludge management programs. One program to increase the digester capacities at two San Antonio, Texas treatment plants involved use of mobile equipment for thickening sludge. Rotary drum dewatering units were set up at each plant to thicken waste activated sludge at .7 to 1.5 percent solids as it came from aeration ranks. Sludge entering the thickener at 200 gallons per minute was dewatered to 5 to 7 percent total solids. Upon leaving the digesters, the sludge was belt filter pressed by additional mobile equipment, which accepts aerobically digested and blended anaerobic and aerobic sludges for dewatering. All the mobile dewatering units were equipped with metering devices and polymer mixing and feed systems.

Bulking Agents

Sludge composting requires use of a bulking agent to improve porosity and keep the compost mass from becoming anaerobic. The sludge solids content directly affects the amount of bulking agent required for the composting process to function properly. A low solids content, for example, means a greater amount of bulking agent will be needed. Increased use of bulking agent directly affects the sizing and capacity of a facility.

Prepared by Dr. Eliot Epstein, the two tables below demonstrate the economic relationship between solids content and bulking agent use. They show the annual cost of a bulking agent for various prices and sludge solids content for a 10 and 25 dry ton sludge facility, respectively. The illustration assumes a 70 percent bulking agent recovery factor. The costs would be higher if less bulking agent is recovered, and lower if more is recovered.

The impact of sludge solids can be readily seen, points out Epstein: For example, for a 10-dry ton facility at a 70 percent bulking agent recovery and $5/cy bulking agent cost, increasing the solids content from 18 percent to 20 percent represents a savings of $13,578 or 16 percent.

If a 10-dry ton facility having vacuum filters and producing 18 percent cake changes to filter belt press equipment producing 26 percent cake, it could realize an annual savings of $43,779 or 52 percent. Thus, the pay back period for the filter belt press could be very short. The other operational costs are also affected since more material needs to be handled when more bulking agent is used for a low solids sludge than for a high solids sludge.

Wood chips, recycled compost, and sawdust are the most common bulking agents used for sludge composting. A variety of other materials have been tested and ultimately used, including, tree trimmings, leaves, bark, fly ash, bedding straw, peanut and rice hulls, air classified refuse, and automobile tires.

The Metropolitan Denver Sewage Disposal District No. 1 built a 100 dry ton/day aerated windrow composting facility that began operating in 1986. Prior to the design of a 10 dry ton/day pilot-scale project, the District tested a wide variety of organic composting amendments. The purpose of the study was to identify potential amendments, and determine their physical and chemical suitability, availability, and cost. While the standards of evaluation relative to availability and cost are specific to the Denver area, the findings with regard to physical and chemical suitability are applicable more widely.

The following amendment products were evaluated: Manure products—cattle and poultry; Wood products—tree trimmings, wood chips, sawdust, and wood shavings; and Agricultural products—straw, spent barley, beet pulp, corn silage, and dried corn stalks. The average amendment characteristics (physical and chemical) are presented in Table 1. (This information was compiled by Richard Kuchenrither, Douglas Smith, and William Martin.)

Based upon these evaluation criteria, wet poultry manure and corn silage were dropped from further consideration due to their low total solids concentrations; dry poultry manure was dropped because of its intense odor. Ultimately, Metro Denver decided to use both wood chips and sawdust as its bulking agent in the composting process.

Shredded Tires: Shredded rubber tires also have been used as a bulking material. The West Windsor Water Pollution Control Facility in Windsor, Ontario was the first sludge composting facility to use shredded rubber tires as a bulking agent on a full-scale basis. The facility also experimented with a combination of shredded tires and wood chips, to enhance the composting process.

Dr. Andrew Higgins, a composting consultant, con-

Annual Cost ($) Of Bulking Agent For Various Prices and Sludge Solids Contents at 70% Recovery For 10-Dry Ton Per Day

Facility Bulking Agent Price $/cy	% Solids 18	20	22	24	26
5	83,778	70,200	59,091	47,667	39,999
6	100,534	84,240	70,909	57,200	47,999
7	117,289	98,280	82,727	66,734	55,999
8	134,045	112,320	94,546	76,267	63,998
9	150,800	126,360	106,364	85,801	61,998

Annual Cost ($) Of Bulking Agent For Various Prices And Sludge Solids Contents At 70% Recovery For 25-Dry Ton Per Day

Facility Bulking Agent Price $/cy	% Solids 18	20	22	24	26
5	209,445	175,500	148,728	119,168	99,998
6	251,334	210,600	177,273	143,001	119,997
7	293,223	245,700	206,819	166,835	139,997
8	335,112	280,800	236,364	190,668	159,996
9	377,001	315,900	265,909	214,502	179,996

ducted research at Rutgers University on the use of shredded rubber tires as a bulking agent. He noted that at the Windsor facility, one of the biggest problems experienced was the unavailability of a high quality product that was both uniform in size and contained cleanly cut edges. One way to overcome this problem is to specify the size and limitations on the amount of over-sized and under-sized rubber chips allowed.

Another problem was that although tire chips produce a good structure for composting, there is poor adhesion to the sludge particles. As a result, says Higgins, the sludge does not stick to the surface of the chip but instead tends to form sludge balls rather than forming a thin layer of sludge around the chip. This results in odor production during the initial stages of composting. However, as the shredded rubber is reused, surface adhesion increases and sludge coats the surface of the chip.

The Windsor facility addressed this problem by mixing wood chips with the shredded rubber. This resulted in increased surface area with the wood chips serving as fines to increase the size distribution of the bulking agent, notes Higgins. Wood chips also decreased the moisture content of the mix, which is important for efficient composting.

The Rutgers experiments were conducted to determine the optimum mix ratio and aeration rate for composting dewatered sewage sludge using shredded rubber tires as the bulking agent. Heavy metals content and available markets and projected costs were also analyzed. The conclusions of the study were as follows, reports Dr. Higgins:

Thermophilic composting can be sustained throughout the recommended 21 day active composting period. Successful composting is highly dependent on the type of sludge used in the mixture and its initial moisture content. The raw primary sludge used in these experiments had an organic matter level of between 70 and 75 percent and an average moisture content of 28 percent. These are optimum levels for successful sewage sludge composting with shredded rubber.

Chip size is important in maintaining adequate aeration. A chip size ranging from 3/4 × 3/4 inch to 2 × 2 inches is recommended. This combination should provide adequate pore space and surface area. The aeration rate is important in maintaining aerobic conditions and in removing excess moisture from the pile. Aeration rates of 1000 cf/hr/dt have been very effective in removing moisture. However, because of the lower amount of energy available using shredded rubber, this high airflow rate may be difficult to maintain during cold weather. In that event, lower airflow rates such as 500 cf/hr/dt should be used.

The heavy metal analysis of the finished compost indicates that increases occur as a result of composting with shredded rubber. Iron and zinc are present in the shredded tires and accumulate in the finished compost. The results of the tests indicate that these metals are slowly released to the compost and will gradually decline as the rubber is recycled. Other metal increases will be modest and should not significantly degrade the quality of the product.

Bedding Straw: In an experiment to find an alternative bulking agent to replace sawdust in its windrow composting program, the Los Angeles County Sanitation Districts tested bedding straw from horse stables at racetracks in southern California. It was estimated that up to 2,500 cubic yards per day of bedding straw could potentially be diverted from landfills and substituted for the sawdust. Of major concern was determination of the effects of the straw on drying rates, internal temperatures, disinfection, odor emissions and final product quality.

Side-by-side experiments showed that straw windrows compared flavorably with sawdust windrows in all categories except odors. The greater thermophilic activity resulted in a more rapid drying rate in the straw windrow. To explain the higher temperatures and faster drying in the straw windrow, it was suggested that biodegradation of bedding straw in the straw windrow during composting liberated more energy (heat) than the biodegradation of sawdust in the sawdust windrow. It was estimated that in the full scale operation the productivity of straw windrows would exceed the productivity of sawdust windrows by over 25%.

Both windrows effectively inactivated bacterial indicators and pathogens.

Concluded Jonathan Hay, Ross Caballero and other researchers in the Los Angeles study: "The straw windrow decreased in volume more than the sawdust windrow. Final compost products were well-stabilized, humus-like in texture, uniform in particle size, and contained no objectionable odors. Because potassium was higher in the straw than in sawdust, the straw compost product contained more potassium than the sawdust compost product. An economic analysis suggested that the daily bulking agent cost would not increase if the Districts' composting operation used straw in lieu of sawdust.

The emission of obnoxious odors was the major drawback of straw windrow composting. Because of the high odor annoyance potential, full scale composting of the straw windrows at the JWPCP or elsewhere where odors would present a public nuisance is not feasible. However, it might be possible to use straw as a substitute for the sawdust in an in-vessel composter. Unlike the windrow process which releases odors directly into the atmosphere, in-vessel systems contain odors within the reactor or reactor building. Odorous air is collected and piped through an odor treatment system before it is discharged into the atmosphere. WHen used in an in-vessel application, bedding straw may prove to be a very cost competitive alternative to sawdust."

Fly Ash: The Scarborough, Maine Sanitary District began operating an aerated static pile composting facility in 1984, using wood chips as a bulking agent. After a few months of operation, however, the following problems arose: 1) A lack of space, because only 10 yards of sludge could be placed on one aeration pipe; 2) Odors; and 3) Infrequent screening because the screen ordered had not been delivered. Scarborough decided to try using fly ash from S.D. Warren, a local paper company, in an effort to address these problems. John Hart, operator of the facility, describes the current operation as follows:

The sludge and ash are mixed in a portable pug mill unit (12 cubic yard SSI Mixer) at a ratio of one to one by volume. Approximately four to five cubic yards of ash and the same amount of sludge are both added to the pug mill and mixed for two to three minutes. The material is then discharged onto the floor and then piled onto a wood chip covered aeration pipe. The chips around the aeration pipe allow for more uniform air transfer and help keep the sludge ash mixture from sifting into the pipe.

After construction, the pile is then covered with a six-to-12-inch layer of ash. The ash serves both as insulation and as a deodorizing blanket.

The pile is then aerated by blowing into the pile 24 hours per day until the pile is broken down. Temperatures of 60° to 70°C are reached within 24 hours and maintained for up to 14 days.

When delivered, the fly ash moisture percentage equals 32 percent. The typical moisture of freshly-mixed sludge and compost at a 1:1 ratio is 50 percent. Analysis of a compost pile built on Aug. 20, 1985 and broken down Sept. 3, 1985, showed the following moisture percentage after 14 days of composting: Bottom of pile—11%; Middle of pile—14%; and Top of pile—42%.

Because there is no need to reclaim or screen the material for a reusable bulking agent, it saves a great deal of time by simply stockpiling the composted mix after seven to 14 days and then marketing the product.

Some of the benefits of using fly ash are as follows:

1. Reduction in odor due to the high carbon content of the ash;
2. Reduction in space as a chip/sludge mixture of 1:1 would cover three to four pipes versus an ash/sludge mixture covering one pipe; and
3. No screening time—the product is marketed "as is".

The only real disadvantage of the fly ash is the dust created when it dries out.

Various innovations have also been tried with more conventional bulking agents, like sawdust and wood chips. Schenectady, New York, for example, installed an aeration system in its amendment (sawdust) storage silo to increase the dry solids content of the material from 50 to 55 percent or greater. This builds in some flexibility to achieve the desired solids content going into the system's reactor. Springettsbury Township, Pennsylvania undertook the vigorous challenge of growing its own bulking agent. During the evaluation phase, the township recognized that wood chips would be a major expense associated with composting. (Estimates were that wood chips purchased from a supplier would cost twice as much as those the township could supply themselves.) The engineer proposed growing their own chips by means of a hybrid poplar tree plantation. In 1981, a 135 acre farm was purchased for that purpose. Seven acres of available park land were used for a nursery. James Crooks, superintendent of the Springettsbury Township Wastewater Treament Facility, gives the following summary on this effort:

"The township's experience with supplying its own wood chips has been found to be very labor-intensive and very costly. . . . Looking back at our experience, we believe that it would have been better to start on a small-

TABLE 1

Average Amendment Characteristics

Item	Bulk Density (lb/cu yd)	Total Solids (%)	Total Volatile Solids (%)	pH	TKN (ug/g)	NH_3 (ug/g)	Cd (ug/g)	Pb (ug/g)	Ni (ug/g)	Zn (ug/g)	Cu (ug/g)
Cattle Manure	1,628	58.9	44.8	6.9	19,700	3,900					
Wet Poultry Manure	1,796	28.9	64.9	7.0	62,800	5,447	0.07	8.4	9.3	360	33
Dry Poultry Manure	879	86.2	59.9	7.0	—	3,390	0.09	10.4	8.3	232	33
Tree Trimmings	272	70.5	79.3	5.5	8,700	110	0.18	4.5	1.5	42.4	5.9
Wood Chips	322	66.9	93.6	4.2	156	<0.3	0.11	0.8	2.3	21.4	2.5
Saw Dust	392	75.6	88.6	4.5	1,186	5	0.11	3.3	2.9	35.8	26.1
Straw	57	86.5	80.3	6.6	5,380	89	0.04	1.7	2.9	6.5	2.3
Spent Barley	452	95.2	71.9	5.2	38,700	37	<0.02	0.5	5.2	91.9	27.4
Beet Pulp	354	79.2	73.5	5.5	14,100	181	0.08	1.0	2.0	9.5	5.8
Corn Silage	479	31.8	91.7	4.6	13,400	735	0.20	2.0	1.0	27.2	5.0
Dried Corn Stalks	34	88.4	79.6	6.0	7,370	200	0.05	2.0	2.0	13.0	3.8
Typical Metro Digested Sludge	1,684	2.8	60.4	7.6	110,000	58,000	36	300	200	1,400	900

scale, pilot basis, instead of undertaking the project on the large scale that we did initially. A pilot of several acres would have demonstrated many of the pitfalls."

Enclosures

Enclosures for compost facilities have been built as a step to control odors as well as to improve process control and public acceptance.

The most common approach to covering a composting operation is to build an opensided building, i.e. to simply put a roof over the area needing cover. Oftentimes, these same structures can be fitted with sidewalls if necessary.

A variety of factors have an impact on a decision to cover and/or enclose all or various components of a composting operation. "Climate and neighbors and public acceptance issues drive facilities to covers," says Richard Kuchenrither, a consultant with Black & Veatch in Denver, Colorado. "Covers are very important for process control, especially if a facility is in a climate with lots of precipitation. There can also be some odor control effect from enclosing buildings because it prevents wind from blowing across the piles, which enable odors to escape. But covers only capture odors. Then a system must be designed to treat them."

A potential drawback to covers, he adds, is that they can shield the piles from the sun, retarding the drying effect. A solution is to install covers with solar panels, which allow 90 to 95 percent of the needed radiation to get through. This design was used effectively at Denver Metro's aerated windrow composting facility.

John Donovan of Camp Dresser McKee in Boston sees a trend toward designing covers for aerated static pile facilities. "For the active composting piles, we are generally using buildings where the sides can be closed if necessary, and air collection and odor control systems can be added," he says. "With curing and storage, we use a shed type of arrangement."

The rationale for covers, he adds, is to control moisture—the most critical aspect of composting—and to insure that weather conditions don't have a severe impact on the operation. "From just purely an operational standpoint, it is best to keep the front end loader operators dry to enable them to work most effectively and to keep outside factors—such as slippery conditions—from affecting them."

The trend toward covering and/or enclosing these operations has resulted in more expensive facilities. "Costs can increase by up to one-third," says Donovan, "especially if odor control is required."

When designing a facility or expanding an existing one, determining how far to go with covers and enclosures often boils down to economics—when the costs of enclosing a non-reactor system begin to approach the cost of an invessel system. "That is very site-specific," says Kuchenrither. "In the case of Denver, and in Philadelphia where the facility has been enlarged, it was less expensive to put on covers and construct buildings than to build an invessel system."

The following are reports on enclosures at various sludge composting facilities:

Montgomery County, Maryland: This facility, designed to handle 400 wet tons per day of sludge, currently receives 200 wet tons per day. When it first began operating, the mixing area and the active composting piles were in open-sided buildings. The screening operation was put in a separate building. The facility is bordered by residential areas and an industrial park. From the beginning, it received odor complaints.

"We did odor emission and odor modeling work and found that when the atmosphere is very stable, low level

Cr (ug/g)	Cr (ug/g)	Fecal Coliforms (mpn/g)	Fecal Streptococci (mpn/g)
		1.6 (10^7)	2.8 (10^7)
9.2	9.2	4.7 (10^7)	1.7 (10^9)
7.9	7.9	—	—
1.7	1.7	—	—
4.2	4.2	—	—
6.0	6.0	—	—
3.1	3.1	—	—
0.5	0.5	—	—
2.5	2.5	—	—
2.0	2.0	—	—
2.0	2.0	—	—
800	800	1 (10^7)	1 (10^8)

odor emissions from the aeration pad would accumulate close to the ground and travel off site very slowly,'' says Charles Murray, chief operator. ''There was no dispersion in the air. That pointed to the need for enclosure.''

The active composting area was enclosed by putting up heavy fabric sides. Dispersion enhancement technology was used to exhaust the air into the atmosphere (the process air goes through a separate odor scrubbing phase). Murray says that a study of odor sources at the facility—measured by surface area—found that about 50 percent come from the surface area of the compost piles. About two percent are from the mixing area and 11 percent from the curing piles (because at that point the woodchips are removed and the pile area is much smaller).

More permanent garage door-like sides were installed to replace the fabric walls. Mechanical mixers are also being installed, and that area will be enclosed as well.

Looking back, Murray concludes that if they knew five years ago what they know now about controlling odors, it may not have been necessary to enclose the process area. ''We believe that 95 percent of odor emissions at this site come from the process air stream. If we had been able to solve that problem we might have been all right.'' Nonetheless, from a political standpoint, Murray says it was best to enclose as much of the operation as possible because of the location of the site (especially since a hotel and restaurant are opening right outside of the facility's gate!). The cost of the original enclosure was $2 million, which included the fabric sidewalls and extending the ends of the building to accommodate the exhaust system. The permanent walls being installed will cost about $1 million.

Philadelphia: This sludge composting facility has been expanded to accommodate up to 400 dry tons per day of sewage sludge. Currently, the facility is composting 175 dry tons per day. As part of the expansion and upgrade, several new buildings were constructed. There are two 64,000 sq. ft. drying sheds, open on all sides, with skylights in the roof. Aeration blowers have been installed to further dry the compost prior to screening. ''These sheds will be a big help, particularly during the rainy seasons when the material would get so wet that screening capacity would decrease,'' says Kate Ellis, assistant manager of operations.

The city never considered covering the active composting piles because the area is too vast. Ellis adds that very few problems are experienced during that phase of composting. ''We can control the process by adjusting aeration rates.''

Mechanical mixers, housed in an enclosed building, just began operating. This should help to reduce odors at the site. ''With front end loaders, mixing was the most odorous part of the operation,'' says Ellis.

Scranton, Pennsylvania: The Scranton Sewer Authority has two curing buildings used to get the material dry prior to screening. Covering the active composting piles would be economically prohibitive, says Alfred Polidori, manager of the authority, who adds that cold weather and rain don't seem to affect the composting process itself. ''Liberal use of wood chips and an insulating blanket keep the piles from getting too wet,'' he says. ''And when the weather is really bad we make sure the piles we want to screen two weeks from now are put under cover and spread out so they can dry for 10 days to two weeks.''

Materials Handling, Screening, And Computerization

Several other areas related to the design of a composting facility include materials handling, screening, and the use of computers for process control.

Materials Handling: Two areas which form the primary basis for design of materials handling systems are: 1) media (material) classification; and 2) conveyor selection. Daniel Kelly and Howard Beissel, consulting engineers who helped to design Cape May, New Jersey's in-vessel composting system, discuss these two areas:

Conveyed media (materials) is categorized in material handling design manuals (published by CEMA—Conveyor Equipment Manufacturers Association) based upon varying properties such as density, moisture content, particle size, flowability, angle of repose, abrasiveness, corrosivity, etc. Because information in the manuals is compiled on the basis of average results of a specific media conveyed in various applications, it is best to test the materials to match the media to the conveyor equipment.

Unlike static pile or windrow methods of composting, where material is generally transported by front end loaders and hauling vehicles in a batch operation, a mechanical in-vessel system operates on a continuous flow process requiring a conveyor system for media transport. Using conveyor systems, material can be metered mechanically and totalized on a volumetric and/or weight basis until the required daily volume of each constituent has been processed. This type of material handling system has the potential of greater control and flexibility. Feed equipment is designed with the capability of varying output rates in order to obtain the proper solids balance given normal and variable material characteristics.

As previously indicated, conveyor selection is dependent on a number of specific applications within the overall system.

Detailed analyses for conveyor selection are time consuming but can be efficiently completed through an organized evaluation of critical design parameters. The following list of design parameters is typical for the evaluation of various conveyor systems:

- properties of conveyed media (weight, volume, density, moisture content);
- load evaluation;
- load path (geometry);
- conveyor function;
- throughput rates;
- sizing, speed and horsepower requirements;
- appurtenances required;
- operation and maintenance costs; and
- capital costs.

While there are two basic categories of materials handling systems—unit and bulk materials—it is the latter that applies to composting facilities. Bulk material handling units used throughout the in-vessel composting industry generally consist of six types:

- Belt conveyors: These are the most common due to their reliability, efficiency and relatively inexpensive cost both in terms of capital and operation. Principal drawbacks are limited incline angles of 35 to 45° maximum (20° for sludge cake); and bulk appurtenances such as protective weather covers, catwalks, safety devices, and in some applications, large support structures. Special purpose belt conveyors are capable of conveying material on an incline up to 85 to 90°.
- Drag chain conveyors: These are typically selected over belt conveyors when steep inclines are required. Despite higher horsepower requirements and moderate to high capital and operating costs, drag chains are compact in cross section, completely enclosed and dust tight, and capable of conveying at virtually any angle or configuration. Also, very few appurtenances are necessary and they have been proven in service applications for over a decade conveying dewatered sludge cake, sawdust and compost at numerous European composting facilities.
- Screw conveyors: These have an excellent record for metering and evacuating hard to handle media under a head of material. Variable pitch flighting should be used for live floor applications to insure consistent material withdrawal across the bin or trough. Removal by this method enables the media above to drop uniformly downward thus virtually eliminating the chance of material bridging in the bin. For in-vessel composting systems, horizontally positioned screw outfeed devices are typically used for evacuating material.
- Live Bottom Hoppers/Bins: Live bottom hoppers and bins are generally used at in-vessel facilities for temporary storage and/or for metering the media. For sludge applications, live bottom units should be designed for use on a short term basis and not as a long term storage structure. A day bin prevents materials with high densities and moisture content from packing or matting under pressure, a possible result with long term storage. A series of screw conveyors is the most reliable type of live bottom bin when handling sludge cake, carbon amendment and compost. However, an alternate live bottom bin system successful in handling sludge cake is a circular table with raking arm and hopper.
- Apron Conveyors: Apron conveyors are widely used throughout the solid waste industry and are used in co-disposal (solid waste/sludge) composting systems. They are usually large, heavy duty units with a base and side panels consisting of bolted steel slats supported on a roller system. Apron conveyors are high in capital cost and are not recommended in applications for wet or sticky media such as sludge cake and would generally not be cost-effective for amendment material or compost product.
- Pneumatic Systems: Pneumatic systems have been used in several European in-vessel applications for loading amendment storage facilities with either dry sawdust or shredded paper using forced air. Schenectady, New York uses a pneumatic system in its amendment silo. Typically, pneumatic conveyance is not used for process material conveyance since metering and dust control are limiting factors.

Metering Systems: The metering of materials may be accomplished solely on a volumetric basis using screws and belt feeder systems. However, metering accuracy is generally greater when material weight is recorded with a weigh scale. In addition, unlike weigh scales, screw metering and belt feeder systems typically operate as a separate, self contained unit.

Mixing Systems: Mixing for mechanical composting processes can be accomplished by one of two basic types of continuous flow mixers. The first is a pug mill mixer consisting of two parallel counter-rotating shafts with paddles fastened to each shaft. As material passes through the mixer, it is compressed and sheared by the rotating action of the paddles, resulting in a reduction in output volume typically of as much as 10 percent.

The second type of mixer consists of a circular barrel with a single shaft. Again, paddles are attached to the shaft to convey material through the mixer. Although considerably higher in capital cost, barrel mixer manufacturers claim the equipment offers the advantage of zero compression of the material.

Screening

The main purpose of screening in a sludge composting operation is to remove as many wood chips as possible from the finished product. Screens are also used in solid waste composting to separate out non-compostable items and refine the material for marketing purposes.

When selecting a screen, it is advised to field test it at the location where it will be used before equipment is purchased. The City of Philadelphia tested a number of screens during the expansion phase of its sludge composting facility. The efficiency of the screens was determined by comparing the following parameters: recovery ratio versus capacity and feed percent solids, and final product quality versus percent total solids of feed. For example, says Jerry Vetter of the City of Philadelphia, the percentage of fines in the wood chips after screening should be less than 10 percent. However, the percentage of fines in the screened compost product should be in the 70 to 80 percent range.

To calculate the cost of screening, Philadelphia analyzed two areas: the cost of operation of the screen including all man and equipment hours and the cost savings due to recovery of wood chips. To compute the savings from wood chip recovery, the city assumed that two-thirds of the chips used in the composting process are recovered from screening.

Vetter points out that with any system, various operational problems will cause screen downtime and reduce the screening capacity at a composting operation. Extensive screen testing prior to purchasing a screen should help to keep operational problems to a minimum.

The main concern for any screening system, he adds, is the percent solids of the feed compost. A feed compost with a solids content less than 55 percent will cause most screening systems to blind which resuts in screen downtime for cleaning. However, the trommel screens will usually not begin blinding until the feed is below 50 percent solids for an extended period.

Vetter also made the following observation with regards to efficiency of a screening system:

From an economics viewpoint, a good screening system can reduce costs in three areas of a composting operation:

1. Due to increases in efficiency of the operation, the cost of operating the screen system decreases.
2. Due to increases in wood chip recovery from screening, the purchase of new wood chips is decreased.
3. Materials handling and storage costs will be decreased due to the volume reduction from unscreened to screened compost.

Operationally, an efficient screening system provides a consistent wood chip supply for bulking, and storage area is required due to the volume reduction.

Regarding the compost utilization programs, an efficient screening operation can produce the following results. Economically, a screened compost is desirable since more sludge is being transported and utilized rather than wood chips. Also, a screened compost is more advantageous to a marketing program since it is a better product from an aesthetic and scientific viewpoint. The removal of the wood chips increases the availability of the nitrogen in the compost which is the limiting factor for plant and vegetative growth. Also, the screened compost is drier and easier to handle than unscreened compost which increases the operational efficiency of any compost application program.

Computer Systems

Use of computers for data collection and analysis and process control is common in many sludge and solid waste composting operations. Aerated static pile systems, as well as in-vessel operations, utilize computers as part of their temperature control feedback system. In Akron, Ohio, for example, temperature controllers were installed within various sections of the Paygro in-vessel system. They control the blowers via a temperature feedback loop to the programmable controllers. Air blowers and air control valves can be instructed by computer signals as well. The system is set up so that air flow can be reversed automatically—to either draw air in or out. Computer instrumentation and control also is used in the metering and mixing portion of the system.

Dr. Harry Hoitink of Ohio's Agricultural Research Center in Wooster, has written computer programs for data collection and process control for a mechanical composting system. The Data Acquisition and Controller Center performs two basic functions: To keep track of the length of time the material has been in different parts of the system; and 2) Track the number of days of exposure to 55°C for kill to pathogens, and once completed, adjust the blowers automatically to obtain desired temperatures. In order to implement this system, temperature monitoring devices are placed throughout the plant. Each device regulates a fan, and the entire process can be controlled by a computer, or directed by programmable controllers.

Some other computer applications at composting facilities include:

1. Determining moisture content for windrows. Leon Chesnin, a waste management specialist in the Agronomy Department at the University of Nebraska-Lincoln, wrote a computer program called "Compost" to assist operators in achieving an initial 60 percent moisture rate in a windrow. The program facilitates the combining of the correct amounts of the two-phase compost mixture components—waste being composted and organic drying bulking agent. It requires information on the amount and solids content of the industrial or municipal slurry or solid waste and also can determine the optimum moisture content for composting mixtures under different climatic conditions.
2. Uniform air distribution in the static pile: Computer programs have been developed to provide air uniformity throughout a static pile by calculating the number of holes needed per section of pipe. Melvin Finstein and colleagues at Rutgers University wrote a computer program which permits the user to determine the number of evenly spaced holes—of a given hole size—necessary to supply a given amount of air to a five foot section of pipe.

More recently, Edward Stentiford in the Department of Civil Engineering at the University of Leeds in England developed a computer system to monitor static pile performance. In the initial phase the microcomputer control system is set up to provide the minimum rate of aeration necessary to maintain aerobic conditions while permitting a rapid temperature rise to the feedback phase. At the end of the process, when heat loss exceeds heat production and the temperature falls, the control system maintains some aeration. The system used takes data from a series of thermocouples sited at various positions within the pile. The data from these thermocouples is used to adjust the aeration rate appropriate to the set temeprature point.

Scranton, Pennsylvania, which has a 25 dry ton/day aerated static pile sludge composting facility, acquired a computer system for the wastewater treatment plant. With a software package developed by the facility's consulting engineers, the plant and compost facility operators input daily process data and display data graphics to assist in process decisions. At the composting facility, individual record files are maintained for each daily cell, including pile construction data and daily temperature, oxygen, and blower control settings. Graphic displays of the data assist the operators in decisions for adjusting blower settings. Some consideration had been given to automated blower control. However, the data from multiple probes have demonstrated that there are significant variations in conditions throughout each cell and from cell to cell. Operator judgment is necessary for process control adjustments.

5

ODOR CONTROL

ODORS have been and continue to be the most perplexing problem faced by composting facility operators, consulting engineers, and regulators. Some facilities have spent millions of dollars correcting problems that they could not predict would occur during the design phase. These investments have resulted in systems that almost completely control odors generated. And more is being learned about what must be done during the design phase to ensure that an adequate odor control system will be installed.

There are several critical steps that must take place immediately, says John Walker of the U.S. Environmental Protection Agency's Office of Municipal Pollution Control:

1) Identify *all* possible sources of odor;

2) Characterize what the odors are or are likely to be; and

3) Provide for treatment of all these sources of odor for each part of a 24-hour period and for each season. Look at all the possible changes in air direction and air stability over that 24-hour period.

With a complete survey of potential odor problems, then and only then can engineers evaluate the appropriate scrubbing and dispersion systems and choose one that gives the best ease of control in their particular situation. "If you know what you have to scrub out, then you know what flow rates and amounts of chemicals you are going to need," says Walker. "You can also determine if the pumps available with a particular system are adequate for the job."

Proper handling of exhaust air is also critical to an operation. Sussex County, New Jersey's aerated static pile composting facility, at one point shut down due to odors, has developed an effective odor control system that includes good handling of exhaust air. The operators mix exhaust air with fresh air pulled out of the buildings, thereby diluting the odors to concentrations that have minimum impact, explains Walker.

Odor control treatments include bio-filters (earth, compost), activated carbon, and chemical scrubbers. Sludge composting facilities have used one or several of these methods with varying degrees of success. The following is a discussion of solutions implemented at several facilities.

Cape May, New Jersey: The 12 to 20 dry ton/day in-vessel composting system using the Purac technology is operated by the Cape May County Municipal Utilities Authority in New Jersey. The facility actually has three odor control systems built around tower scrubbers in series, one using an acid wash to remove ammonia and other nitrogen amines, and the other a hypochlorite wash to remove the hydrogen sulfides. The system is removing 99 percent of the ammonia and other nitrogen compounds, and 90 percent of the hydrogen sulfides. The only odor gremlin that remains is a process odor from the reactor. "We have had exhaustive tests done, and the results are that there is no single source of odor that is anywhere near levels that could be offensive,'' says William Cathcart of the Cape May MUA. ''It just seems to be that the combination of many minor odors coming from both anaerobic and aerobic sources is enough to be perceived as a nuisance by some.''

To cool gases before scrubbing, Cape May was using a condenser, but has discontinued doing so for the most part. Operators found that blowing air directly into the gases worked just as well. By using air from places in the plant where there are occasional spot odors, they in effect solved two problems at once.

Clinton County, New York: This 20 to 30 tpd sludge composting facility utilizes the Fairfield digester circular agitated bed system. It began operations in May 1986, composting raw sludge, and is designed for 14-day retention in the digesters. The odor control system includes an air collection system attached to the dome of the digesters and a single stage scrubbing tower. There are louvers on the sides of the digester to bring in outside air for the purpose of diluting the process air in the digester. Originally, sodium hypochlorite and caustic soda were used in the scrubbers. The inability of that mixture to treat the organic odors generated led Clinton County to switch to a solution of sulfuric acid and Deamine, a proprietary chemical supplied by Nu-Tech.

As part of a program to improve odor control, three general approaches have been taken: (1) a two stage scrubber; (2) soil filter; and (3) increasing dispersion of process air. The facility has been able to achieve about a 90 percent odor removal rate with its mixture of acid and Deamine.

''With the rain and cold temperatures in northern New York, in-vessel composting seemed to make sense,'' says George Miller, City Engineer. ''And we have had a long and difficult history of odor control problems, both at the wastewater treatment plant and at the sludge lagoons. We knew it would be difficult to get public acceptance of any project, but we determined that in-vessel composting might be a more acceptable option.''

Better odor conditions in 1988 were also attributed to the weather—more air turbulence prevented long calms. In addition, Miller feels that citizen complaints have been minimized because of an understanding that the composting operation is aggressively tackling odor generation. ''We've kept citizens well enough informed that they have become tolerant and have confidence in what we are doing.''

Montgomery County, Maryland: The aerated static pile facility in Montgomery County has been doing what is probably the most in-depth research on odor control to date. The facility is located in a residential area, where the citizens are well-organized as ''odor watch-dogs.'' The pressure has been intense, therefore, to keep odor generation to an absolute minimum.

The facility has made maximum use of controlling odors by maintaining optimum temperature and oxygen parameters in the aerated piles. Studies conducted prior to the facility's start-up in 1983 confirmed the importance of high aeration rates to maintain high oxygen levels and further defined the importance of limiting pile temperature for odor control. These theories formed the basis for many of the standard operating procedures initially. Since that time, the facility has experimented with various aeration rates and pile temperatures to maximize control over odor generation.

Charles Murray, chief operator of the composting facility, describes some of the research and solutions:

Exhaust Air Treatment—The original exhaust air handling system at the MCCF filtered the air through two large (100 feet × 35 feet) compost filter piles. Up to 65,000 cfm of saturated exhaust air filtered through the compost piles at an average temperature of about 55°C. From the start of operations, it was obvious to the operators that this filtration system did not provide adequate treatment of the exhaust air. Samples of the exhaust air before and after compost filtration showed no effective removal of organic or inorganic sulfides. A series of efforts were initiated to improve the odor removal treatment of this exhaust air. The first was an attempt to chemically oxidize certain odor constituents using hydrogen peroxide.

Initial tests were conducted by fogging various volumes of 35 percent hydrogen peroxide into the exhaust air stream. The degree of odor reduction was measured by sampling the air stream and performing odor panel tests on the samples. (The odor panel test establishes an ED_{50}, or ''effective dilution—50%.*'' The ED_{50} number corresponds to the number of volumetric dilutions, with odor free air, which are necessary to reduce the odor level of the sample to that level at which 50 percent of those on the odor panel do not detect an odor. A high ED_{50} would be more odorous and pervasive than a low ED_{50}.)

The use of hydrogen peroxide did reduce the level of odor in the exhaust air by about 60 percent. This initial success at odor reduction with chemical treatment led the facility to further investigate available systems for chemical oxidation of odorous air streams.

In reviewing available chemical treatment systems, one patented chemical odor scrubbing system (made by Quad Environmental Technologies) appeared to be capable of treating large quantities of air without introducing significant back pressure on the air collection system. The system also appeared to be simple to operate and maintain. Operators at the MCCF decided to conduct pilot scale testing of the system for the exhaust air. A two-stage process was proposed; a sulfuric acid stage for solubilization and removal of ammonia and a sodium hypochlorite stage for oxidation of organic sulfides, aldehydes and other odor constituents. The system demonstrated odor removal efficiencies of greater than 90 percent. Based on the results of the pilot scale tests, the WSSC decided to construct a full scale system for exhaust air treatment, designed to treat up to 65,000 cfm of exhaust air.

Initial performance testing of the full scale system had shown odor removal efficiencies in the 95 percent range, based on ED_{50} results. However, operators had difficulty in establishing control parameters to maintain consistently high odor removal efficiency. Techniques for controlling chemical feed rates, which were recommended by the manufacturer, were not sufficient to ensure odor control consistency.

The Montgomery County Composting Facility, as of mid-1988, was continuing to conduct experiments in an effort to maximize the effectiveness of the odor control system. Methyl disulfides appear to be the most difficult odor constituent to remove. Acid plays a major role in the chemical solution going into the scrubber (versus bleach), and the latest experiments revolve around use of peroxide.

The Maryland operation, as part of its odor control work, has done extensive work to determine odor emission rates and lay out control options. Charles Murray describes these efforts (one of which led to the enclosure of the composting pad described earlier):

Odor Emission Rates: Efforts to quantify odor emission rates for each phase of the operation were conducted concurrently with the efforts to optimize operation of the exhaust air odor control system.

Surface Odor Emission Rates were established for various phases of the operation. Estimates were derived by placing a sampling box on the pile surface and circulating a steady rate of odor free air through the box. The air was then sampled as it existed in the sampling box and the samples were analyzed for ED_{50} and butanol equivalent. Butanol equivalent analysis establishes the equivalent concentration of butanol (in mg/m^3) which gives odor equal in intensity to the sample. The butanol equivalent measurement is a way of standardizing the odors to establish emission rates which can be mathematically manipulated and modeled.

The exhaust air, even treated, represents by far the major source of odor emissions, due to the high volume of air being discharged. The active compost piles represent the next greatest source of emissions. Based on the data, these two sources represent about 95 percent of the total odor emissions. Subsequent modeling of off-site odor impacts and development of control options focused primarily on these two sources.

Control Options: The Industrial Source Complex Short Term (ISCST) model, developed by the Environmental Protection Agency was used to evaluate odor impact on the community. Details of the modeling efforts are described in Study Reports by Greeley and Hansen, January 1986* and Dames and Moore, November 1985.* Ten specific community receptor sites were identified for evaluation and, for selected alternatives, the maximum odor concentration (in butanol equivalent concentration) at any potential receptor location was calculated. The model was used to characterize odor impacts from the operation as it existed, as well as for a series of control options for both the exhaust air and compost pile or area sources.

For the treated exhaust air, several dispersion enhancement alternatives were evaluated: 1) increased discharge stack height; 2) increased air volume flow rate (dilution); and 3) increased air discharge velocity. A total of 18 control alternatives, ranging from 65 to 110 feet for stack height, 35,000 cfm to 500,000 cfm in volume flow rate, and discharge velocities from 30 to 80 feet per second were evaluated. The most effective of these alternatives was exhaust air dilution to 500,000 cfm and discharge at a height of 110 feet and velocity of 54 feet per second. The predicted maximum concentration was reduced from about 500 ug/m^3 butanol equivalent to 32 ug/m^3. This exhaust air dilution/dispersion alternative was the most effective of those evaluated with the model.

An exhaust air dilution/dispersion system was designed and constructed in early 1986. A 14 foot diameter fan was mounted over the scrubber exhaust stack to collect the exhaust air, dilute it to 500,000 cfm and discharge it vertically through a 14 foot diameter stack, which extends to an elevation of 110 feet. Early indications are that this system is performing as predicted and that odor impacts from the exhaust air have been greatly reduced.

The modeling results for the area sources (compost piles) indicated that without additional controls, off-site impacts from this source would be as prominent as those from the exhaust air source. During certain atmospheric conditions (stable atmosphere, low wind speed) natural dispersion of these area source emissions was minimal, and predicted off-site concentrations of odor were sufficient to be a problem. Several alternatives for dispersion enhancement for control of the compost pile surface odor emissions (area sources) were evaluated. Perimeter walls of varying height and several pile enclosure options were modeled for impact on community receptors. One of the most effective of the area source control options was enclosure of Pads 1 and 2 (where piles are aerated) with discharge of air ventilated from the enclosure from a 14 foot diameter roof mounted fan.

**Source Emission Odor Measurement by a Dynamic Forced-Choice Triangle Olfactometer. A. Dravnieks and W.H. Prokopy. Air Pollution Control Association Paper No. 73-276. June 1973.*

**Odor Control Study of the Montgomery County Composting Facility. Greeley and Hansen. January 1986.*

**Evaluation of Odor Control Alternatives for the Montgomery County Composting Facility. Dames and Moore. November 1, 1985.*

Delaware Reclamation Project

The following report on research to develop an optimum system for odor control for an in-vessel facility was prepared by Pasquale S. Canzano and Kenneth J. Aiani. Mr. Canzano is Chief of Engineering and Operations for the Delaware Solid Waste Authority and Mr. Aiani is General Manager for the Fairfield Service Company at the composting project.

The Delaware Reclamation Project (DRP) in New Castle County, Delaware simultaneously processes 1,000 tons per day of solid waste and up to 350 tons per day of digested sewage sludge into recycled products. The DRP products include a refuse derived fuel, ferrous metals, glass, non-ferrous metals, and humus. The subsystem that produces the humus product utilizes aerobic digestion in a closed composting vessel. The sewage sludge is mixed with a solid waste substrate in a dry weight ratio of approximately 1 to 3.4. The sewage sludge is delivered to the DRP by the City of Wilmington from the City's wastewater treatment plant. Prior to delivery, sludge is anaerobically digested, then dewatered to a solids content in the range of 18 to 22 percent using belt filter presses. The solid waste substrate consists of the organic components of the solid waste after the refuse derived fuel, ferrous metals, non-ferrous metals, and glass products have been removed.

Equipment & Controls

The DRP has four closed vessel aerobic digesters, each designed to process a mixture of 175 wet tons per day of sewage sludge and solid waste substrate at the specified mix ratio. The biomass within the digesters is subjected to a five-day residence time at a temperature in the range of 140°F to 160°F. The digesters are equipped with staggered augers on a rotating bridge, which distribute and mix the biomass within the digesters while continuously discharging humus. The biomass is continuously aerated to allow biological decomposition by in-situ thermophyllic bacteria. Oxygen (O_2) levels are monitored in each of digesters by continuous on-line oxygen analyzer/recorders. Aeration is automatically controlled and the system is designed to maintain uniform temperature and O_2 concentration profiles across the digester. The O_2 concentrations typically are maintained in the range of 12 to 17 percent. Each digester has a blower and dome ventilation fan to control aeration and temperature. Currently, the four digesters have a daily production rate of 300 tons per day of raw humus product or 75,000 tons per year. The DRP also has the capability to screen and refine the raw humus to produce high quality granulated and/or pelletized humus products for agricultural and horticultural applications.

Operating Problems

The operation of the aerobic digesters has encountered certain problems intermittently. A major one was nuisance odors that were annoying to the local community when certain wind conditions prevailed. In addition, certain physical and electrical problems occurred depending on the digesters' mode of operation. These problems included water vapor formation and heat rise in the digester domes, poor aeration of the biomass, corrosion, and electrical short circuits from condensing moisture.

Digester Operating Modes

Three different digester operating modes were evaluated in a developmental program to improve the digester operations with emphasis on controlling the odors.

A. Mode 1 Operation

In Mode 1 operation, air was sparged through the biomass using forced draft fans and a concentric series of perforated plastic pipes at the bottom of the digester. After passing through the biomass, the air exited the digester through the dome ventilation fan opening and the annulus between the top of the digester vessel wall and its dome cover. This method did not allow a single point of odor control.

Mode 1 operation caused a slightly positive pressure within the digester and was intended to eliminate odors from the digesters by creating a highly aerobic condition within the biomass. As shown in Figure 1, the dome ventilation fan at location 3 did not meet the vendor's specifi-

Figure 1 **Mode 1 Digester Operating Conditions**

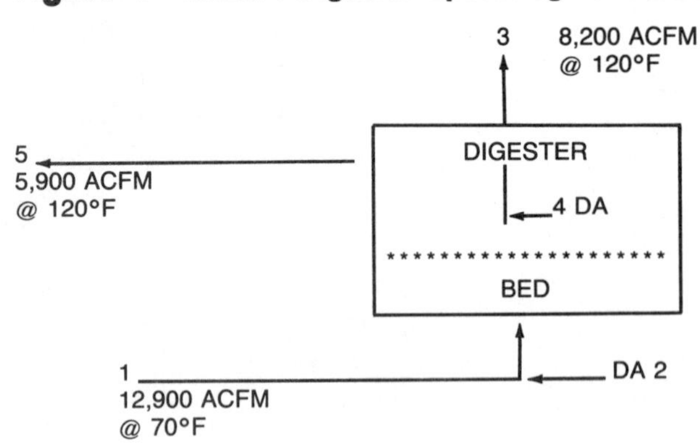

Operation

Case	*Digester*	*Treatment*	*Feed W/Disch.*	*Date*	*Avg. Odor Levels (D/T)*
1-1	B	T1	X	8/18/86-8/19/86	25,000
1-2		T2	X	8/18/86-8/19/86	3,815

TABLE 1.
Digester Treatments

Treatment	Chemical Rate(s)*
T1	VI
T2	V & VI
T3	VII & VIII
T4	I
T5	II & VI
T6	III & VI
T7	IV

**See Table 2*

cations to adequately exhaust the digester air and caused sufficient back pressure to force air out of location 5, the dome annulus. While this mode of operation fluidized the biomass and provided very good air distribution within the digester, it produced excessive water vapor in the dome space creating condensation within the structure, equipment and walkways. The condensation caused corrosion and occasional electrical shorting problems within the equipment. The water vapor rising from the fluidized biomass also created a visibility problem. On cool mornings and evenings it became impossible to see within the digester dome and to perform certain daily maintenance activities.

A 3.0 hp ventilation fan located in the center of the dome cover was operated continuously to remove moisture and heat from the digester's dome, however, it had insufficient air handling capacity for adequate removal. At times, dome temperatures were as high as 125°F.

B. Mode 2 Operation

In Mode 2 operation (See Figure 2), a variety of equipment changes were implemented so that air was drawn down through the biomass from the top of the digester, collected in a plenum, then exhausted through a single vent pipe attached to a chemical scrubber. This mode of operation provided a slightly negative pressure within the digester and was intended to prevent any fugitive odors within the digester dome from exiting through the top. Normally, the ventilation fan was off and was operated only during periods of maintenance within the dome.

C. Mode 3 Operation

In Mode 3 operation (See Figure 3), air was sparged through the biomass in the same fashion as Mode 1, however major changes were made to the dome ventilation fan and the manner in which the digester was operated. The fan was changed from a 48 inch centrifugal roof exhauster to a 60 inch vane axial flow fan and the horsepower of the motor was increased from 3.0 hp to 15 hp. As shown in Figure 3, this change resulted in an air flow of 60,000 ACFM or about eight times as much air flow through the digester dome exhaust at location 3 when compared to Mode 1 operation. In addition, the ventilation fan was always in operation and provided an upward instead of a downward discharge normally associated with a typical centrifugal roof exhauster. These major changes, while substantially increasing the air flow within the digester's dome space, also dramatically improved the digester's performance by realizing all of the advantages of Mode 1 and Mode 2 operations (primarily enhanced air distribution within the biomass and a single-point exhaust discharge). The need for a chemical scrubber was eliminated. Mode 3 operation also caused a noticeable negative pressure within the digester dome space because of the greater air flow rate achieved with the modification of the dome ventilation fan.

Although Mode 3 operation eliminated the need for a chemical scrubber, it allowed for the introduction of a chemical odor neutralizer into the air fluidizing the biomass at location 2 and/or the air exiting the dome ventilation fan at location 4. Introducing a chemical odor neutralizer at these specific points provided additional flexibility to control and minimize odors in the event that upset conditions occurred in the digesters. The chemical odor neutralizer which is currently being utilized is identified by the manufacturer as DeAmine (DA).

Effectiveness of Each Mode of Operation on Reducing Odors

Mode of Operation	Digester Treatment	Odor Level Dilution to Threshold (D/T)	Reduction Over Previous Mode	Total Cumulative Reduction (%)
1	T-1	25,000	0	0
1	T-2	3,815	86%[1]	86%
2	T-3	1,955	49%	92%
2	T-7	1,970	48%	92%
3	T-5	538[2]	72%	98%
3	T-4	567	71%	98%

[1]Compared to T1 @ 25,000 D/T
[2]Average includes one unstable digester (digester D)
Source: The above data has been compiled from reports prepared by Prokop Enviro Consulting, dated September 2, 1986 and TRC Environmental Consultants, Inc., dated October 8, 1987.

Figure 2 **Mode 2 Digester Operating Conditions**

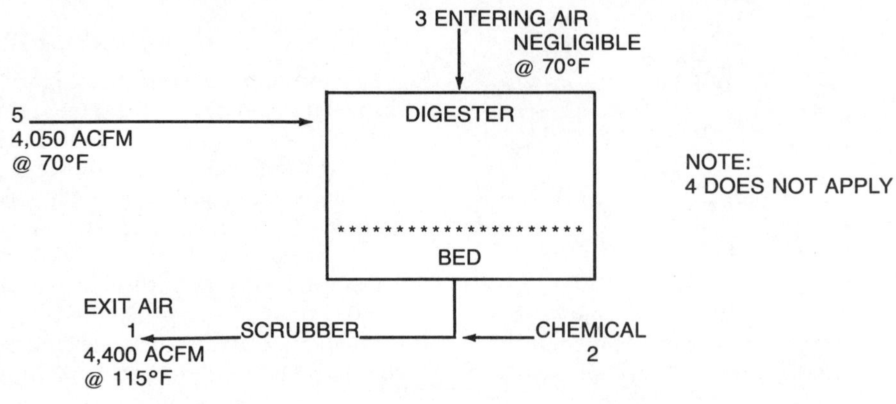

Operation

Case	*Digester*	*Treatment*	*Feed W/Disch.*	*Date*	*Avg. Odor Levels (D/T)*
2-1	D	T4	X	8/18/86-8/19/86	7,555
2-2	D	T7	X	8/18/86-8/19/86	1,970
2-3	C	T4	X	8/18/86-8/19/86	61,750
2-4	C	T3	X	8/18/86-8/19/86	1,955

Methodology

The digesters were operated in the three modes described above in an attempt to improve the physical operating conditions and more importantly to minimize and/or eliminate odors from the digesters. In all three modes,the humus product quality was acceptable and consistent with contract product specifications. Therefore, evaluations were conducted to determine the optimal mode of operation as measured by the level of odor produced during each mode.

During the implementation of Mode 2, the state of the art technology for chemical scrubbers and commercially available odor abatement systems was reviewed in an effort to identify an effective system for treating odors in the digester discharge air. This technology included (a) packed columns, (b) co-current and countercurrent spray chambers, (c) bio-filters, and (d) fluid bed scrubbers. A laboratory program also was initiated to evaluate the efficacy of a variety of chemicals to neutralize the odor.

Unsatisfactory results were obtained during the field testing of a packed scrubber pilot plant unit. However, successful in-house laboratory tests using a combination of sodium hypochlorite (NaOCl) and potassium permanganate ($KMnO_4$) led to the development of a prototype direct contact scrubber, which effectively reduced odors from the digester discharge air when operated in Mode 2. In addition to chemicals, cedar wood chips and limestone in a combination filter were also evaluated to control odors. This simple filter was effective in reducing odors, but after approximately one week, it noticeably began to diminish in its capacity to fully control the odors from the digesters. Conversely, the batch chemical scrubber required purging and chemical recharge after about one month's operation.

Based upon evaluation of Mode 2 operation for controlling odors, utilizing the prototype batch scrubber, a determination was made to proceed with Mode 3 operation to further improve upon the level of odors emitted, and eliminate certain problems being experienced in previous modes. In addition to increasing the capacity of the ventilation fan, modifications were made to other mechanical equipment and means of controlling the process variables to optimize digestion of the biomass and thereby minimize the generation of odors. A discussion of the effects of these modifications on odor control is beyond the scope of this paper.

Mode 3 results were further evaluated using the Gaussian Dispersion "puff" Model. A very conservative approach was taken in determining the data to be entered into the model. The nearest local community receptor residence is approximately one-half mile from the odor source. The model, however, will predict the maximum odor impact, assuming a full easterly wind, for a receptor 2,000 feet downwind. Other "puff" model assumptions incorporating the digester treatment rates shown in Table 1 are as follows: (1) Two of the digesters operating normally with treatment T-5, (2) One digester operating un-

Figure 3 **Mode 3 Digester Operating Conditions**

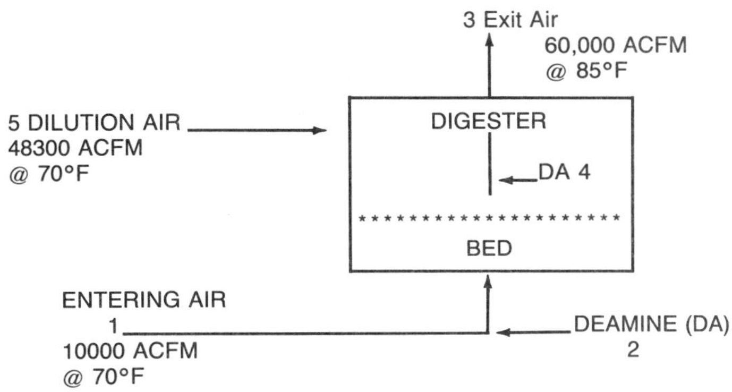

Operation

Case	*Digester*	*Treatment*	*Feed W/Disch.*	*Date*	*Avg. Odor Levels (D/T)*
3-1	B	T1	X	7/27/87	722
3-2		T5	X		499
3-3		T6	X		659
3-4	C	T1	X	7/27/87	625
3-5		T5	X		448
3-6		T6	X		785
3-17		T4	X	7/29/87	567
3-7	A	T1	X	7/28/87	610*
3-8		T5	X		479
3-9		T6	X		473
3-10	D	T1	X	7/28/87	910
3-11		T5	X		725
3-12		T6	X		1083

**Omitting an anomalously low odor value*

der upset conditions with treatment T-5, being only half effective, (3) One digester operating normally with treatment T-4, which provided no odor treatment and (4) Accounting for other measured sources of untreated odor in the vicinity of the digesters. The results of this modeling are summarized in Table 3.

The weather conditions which frequently occur in the area are also listed in Table 3. These conditions are defined as: (1) W-1: unstable meterological conditions with light winds of 1.0 m/s; (2) W-2: very stable conditions (inversion) with light-moderate winds of 2.5 m/s; and (3) W-3: all other meterological conditions.

Dose-response curves were developed using Steven's Law based on odor character and intensity to determine the dilution to threshold (D/T) level at the receptor, that would produce an objectionable odor based on a standard value of 3.5 on the Butanol Scale. This level was determined to be greater than 7 D/T at the receptor, where 1 D/T is the threshold level. The details for this procedure are fully described in ASTM test E 544-75 (1981).

Test Results

Odor measurement and evaluation studies were performed on two separate occasions by independent odor consultants to measure the effects of each mode of operation on odor emissions.

Treatment T2 produced an 86 percent reduction in discharged odors compared to treatment T1 for Mode 1 operation. Although this initially appeared to be quite successful, the odor level of DA treated exhaust air still contained an odor level of 3,815 times threshold—clearly an unacceptable level. An investigation of odor character, however, revealed very high levels of ammonia, ranging from 150-400 ppm, due to unacceptably high temperatures of 160°F to 170°F within the biomass. These temperatures resulted from the accumulation of excessive amounts of heat in the digester dome space, which prevented sufficient removal of heat from the biomass. The high levels of ammonia substantially contributed to the odor levels being measured.

Mode 2 operation included the use of chemical scrubbers, and demonstrated an even greater ability to reduce the levels of discharged odors. However, long term operation produced negative effects such as leachate accumulation and plugging of the air distribution system. This caused anaerobiosis and a more pungent, acrid odor character.

A 70 percent reduction in the level of odors in the untreated air (Treatment T-4) resulted solely in changing from Mode 1 (Case 1-1) to Mode 2 (Case 2-1) operation. Case 1-1 was used as a comparison, since treatment of the biomass with DA was proven to be ineffective in reducing odors. The significant reduction was attributed to improved temperature control resulting from process and equipment modifications, referred to earlier, for Mode 2 operation. Treatment T-7 (Case 2-2), which used DA as a reagent in the scrubber, caused an additional 74 percent reduction in the odor level from a pretreated D/T level of 7,555 to 1,970. This produced a net total reduction of 92 percent for Mode 2 (Case 2-2) when compared to Mode 1 (Case 1-1).

Treatment T3 using sodium hypochlorite (NaOCl) and potassium permanganate ($KMnO_4$) as reagents in the scrubber resulted in a 97 percent reduction in measured odors, from a D/T level of 61,750 to 1,955. The high D/T level of 61,750 for Case 2-2 was attributed to the anaerobiosis resulting from the long term negative effects of Mode 2 operation. Normal Mode 2 operation, however, produced a D/T level of 7,555 (Case 2-1) and Treatment T-7 using DA was as effective as Treatment T3 in controlling the level of odors discharged. Treatment with DA, which used less chemical reagent(s) and allowed simpler treatment operations, was preferred over the NaOCl and $KMnO_4$. In addition all of the major operational problems had been resolved during Mode 2 operations.

Mode 3 (the current mode of operation shown in Figure 3), when compared to Mode 2, resulted in an even greater odor reduction in both the untreated air and the exhaust air discharged from all digesters. Three major variables were evaluated during this mode: (1) the effectiveness of two different points of treatment, (2) the effects of feeding fresh material into the digester, and (3) the effects of a totally quiescent state within the biomass.

The first point of treatment, Location 2, involved the addition of DA into the intake air plenum. This treatment was evaluated using "C" digester. By comparing the results of Treatments T-1 and T-4 (no treatment) in Figure 3, it can be concluded that there was essentially no effect on reducing odor as a result of the introduction of DA into the biomass. A D/T level of 567 without the DA treatment can be compared to a D/T level of 625 with the DA treatment. There also was no significant difference in the D/T levels during feeding and quiescent periods.

The second point of treatment at Location 4, however, did demonstrate a 20 to 31 percent reduction (see Figure 3) in discharged odors with the introduction of DA into the dome exhaust utilizing Chemical Application Rate II. An increase in chemical rate was also evaluated and found to be counterproductive at Chemical Application Rate III. In all cases, as the DA rate increased from 0.08 to 0.33 GPD, the treatment effectiveness declined, as demonstrated by the increased odor levels. This over-application resulted in the detection of vinegar odors associated with the unreacted DA using Rate III, while the discharged odor character using DA Rate II was a typical, treated compost odor. It is also important to note that these results were achieved despite the fact that one digester had not yet fully stabilized in changing from Mode 2 to Mode 3 operation. As shown in Figure 3, digester D had noticeably higher odor levels compared to digesters A, B, and C.

The peak receptor D/T is the maximum instantaneous odor concentration, similar to the method of detection of the human nose and allows a very conservative odor evaluation. Based upon the modeling results for Mode 3 operation, conditions W-1 and W-2 would produce an objectionable odor 0.25 percent of the time only if a digester were experiencing unstable, upset conditions. During Condition W3, which occurs the remaining 99.5 percent of the time, there would be no objectionable odors detected even with one upset digester. Under all three conditions, the *average level of odor at the receptor* would be below the threshold level of 1 D/T.

Summary

Mode 3 operation was the best of the three modes of operation tested and evaluated. The table summarizes the effectiveness of each mode on reducing the level of odors, making it quite clear that there was a substantial and continuous decrease in the levels of odor progressing from Mode 1 to Mode 3 operation. A 98 percent total reduction in discharged odor was achieved using Mode 3, and the operational problems experienced during previous modes were generally resolved. The chemical neutralizer, DeAmine, was most effective in reducing odors at a treatment rate of 0.08 GPD; best results were obtained treating only the air being exhausted from the digester dome at location 4. No significant difference between the odor levels of a quiescent biomass and a mixing biomass were obtained, suggesting that odors are primarily determined by the efficiency of the digestion process and not mechanical mixing, feeding of the digester, or discharge of the humus product.

Soil Deodorization Systems: A number of composting facilities use piles of finished compost to filter exhaust air through prior to release into the atmosphere. Others have had success with soil filters. Mechanisms associated with the soil matrix are physical adsorption, chemical oxidation, and biological assimilation of the material once it has adhered to the soil particle, says Mark Selby, a civil engineer who evaluated a soil filter system at a Florida wastewater treatment plant. He explains:

Two major advantages associated with the biological adsorption mechanism common to the soil filter are:

- Once at a steady state, the biological process is self-sustaining as long as waste loading remains constant and proper environmental parameters—temperature, oxygen and carbon dioxide concentrations, carbon/nitrogen ratio, and others—are maintained.
- Properly designed, the biological air scrubbing process does not require external mechanical or thermal energy other than the energy associated with the blower.

Physical adsorption occurs as the odorous compounds are stripped from the air stream as it moves up and through the soil matrix. The cation exchange capacity (CEC) is used to assess the adsorption capability of a particular soil. Finished compost, humus and peat are being tested as the filtering agent. CEC's typically range between 100 and 180 for that type of media, compared to clay, which drops down to about 20 to 30 units.

Chemical oxidation and biological assimilation are directly related to the moisture content of the soil matrix. A water supply system incorporated into the filter controls the moisture content in the media to maintain hydrogen sulfide removal efficiency of better than 98 percent. For optimum performance, the moisture content should be kept in excess of 55 to 60 percent.

Chemical Treatments: Some composting facilities, as noted above, have developed solutions that make use of commercial products such as Odormaster II from Pepcon Systems or the Quad Chemtact System. According to Quad's Bert deVries who invented the process, his system uses a very small droplet size to wet-scrub odors out of gases—practically a fogging action. DeVries says the key to successful scrubbing is the temperature of the gases.

"Composting generates high temperatures and of course, that's beneficial from the standpoint of killing pathogens," he explains. "But high temperatures cause a long list of organic solvents and chemicals to be released into the saturated vapor. Wet scrubbing action dissolves these compounds into the water phase of the scrubbing solution, but high temperatures often *prevent* adequate solubility. The solution to this problem is to cool the gases down to 80 or 90°F before running them through the scrubber."

Referring to himself as an "odor guru" occasionally, deVries continues: "I expect science to creep into the arguments about odor control any day now," he likes to say, "but sometimes I wonder. Plant managers come to me and say they want 100 percent odor control. There is no such thing as 100 percent odor control. Einstein taught us matter is indestructible. So we can, for example, scrub ammonia out of the gases. But it's still in the water. Plant managers don't want that. They want the ammonia to disappear."

The exhaust air after scrubbing may contain 5 percent or less of the original odor, but even that much can be a problem, deVries notes. "We have numerous ways to handle exhaust air to reduce the remaining odor to where it is no longer a nuisance. For instance, we can dilute it with fresh air or we can scrub it further with plain water."

Compost facilities at the Metro Denver Sewage Disposal plant and Nevada's Clark County AWT plant have reported effective results in removing odorous compounds from air streams with the Odormaster II pilot unit. The Pepcon System uses two scrubber towers and an electrolytic hypochloride generator to provide treatment through absorption, neutralization and oxidation.

The first tower, using either potable water or final plant effluent as the scrubbing solution, cools the air stream and absorbs and/or neutralizes water soluble odorous compounds. Acid may be added to reduce feedwater requirements. The blowdown rate is adjustable for maximum odor removal at this stage.

The second tower uses sodium hypochlorite generated onsite, combined with anodic oxidation, to absorb, neutralize and oxidize the remaining malodorous compounds. This tower employs a hypochlorite generator, utilizing coarse salt, with automatic control to match oxidant production to malodorous gas demand. The chemical reactions involved in treating the contaminants with $NaOCl$ result in the recovery of the salt which is then recycled through the system for reoxidation to $NaOCl$.

6

COMPOST PROCESS DESIGN CRITERIA

Roger T. Haug

(Roger T. Haug, a consulting engineer in Torrance, California, wrote a four-part series of articles under the general title, "Composting Process Design Criteria." Part I covered feed conditioning; Part II covered detention time; Part III covered aeration requirements; and Part IV presented a case study applying the recommended design approach and design criteria to an actual system. This series appeared in BioCycle, *from August through December, 1986. Parts I, II, and III are excerpted in this chapter.)*

COMPOSTING systems are often divided into a first-stage, high-rate phase and a second-stage, curing phase. No precise definition or distinction between these two phases exists. The high-rate phase may use windrow, aerated static pile, or reactor processes. The first-stage is characterized by high oxygen uptake rates, thermophilic temperatures, high biodegradable volatile solids (BVS) reductions, and higher odor potential. The second-stage, curing phase may use aerated or non-aerated piles, windrows, or even enclosed reactors. The curing phase is characterized by lower temperatures, reduced oxygen uptake rates, and lower odor production potential. The curing phase provides the time required for (1) degradation of the more refractory organics and (2) overcoming the "slowing" effects imposed by kinetic rate limitations.

Traditionally, the first-stage of composting has been more engineered and controlled due to the need to reduce odors, supply high aeration rates, and maintain process control. The curing phase is often less engineered and less controlled. While this may be acceptable from a process engineering standpoint, both phases are integral to the design and operation of a complete composting *system*. A process diagram for a generalized composting system is presented in Figure 1.

Nomenclature

X_c = wet weight of dewatered sludge produced per day.
Xp = wet weight of compost product produced per day.
X_r = wet weight of material recycle per day.
X_a = wet weight of organic amendment, other than sludge cake or compost recycle, added to mixture per day.
X_b = wet weight of bulking agent added to mixture per day.
X_m = wet weight of mixed feed materials entering the composting process per day.
S_c = fractional solids content of dewatered sludge cake.
Sp = fractional solids content of compost product.
S_r = fractional solids content of recycle material.
S_a = fractional solids content of amendment.
S_b = fractional solids content of bulking agent.
S_m = fractional solids content of the feed mixture.
V_c = volatile solids content of sludge cake, fraction of dry solids.
Vp = volatile solids content of compost product, fraction of dry solids.

V_r = volatile solids content of recycle material, fraction of dry solids.
V_a = volatile solids content of amendment, fraction of dry solids.
V_b = volatile solids content of bulking agent, fraction of dry solids.
V_m = volatile solids content of feed mixture, fraction of dry solids.

Requirements for Feed Conditioning

Designers and operators of composting systems have a limited number of areas over which to exercise control of the process. One area of control is the composition of the infeed mixture. The operator must adjust the proportions of feed components to satisfy the energy balance and avoid rate limitations caused by lack of moisture, lack of free air space, sterile feed, or low nutrient levels. The designer must know the quantities of required feed components to size the system and its conveyance equipment.

Three approaches to feed conditioning have been used in practice:

1. Recycle of compost product, or material after the first-stage, and blending with the dewatered cake.
2. Addition of organic or inorganic amendments to the sludge.
3. Addition of a bulking agent, such as wood chips, and subsequent screening of the bulking particles from the compost product.

Inputs of sludge cake, amendment bulking agent, and recycle material are shown in Figure 1. This is not meant to imply that all additives would be used at the same time. Whether one or more are used depends on the composting system under study.

Amendments and Bulking Agents

An *amendment* is a material added to the sludge cake to condition the feed mixture. Two types of amendments can be defined as follows:

1. Structural or drying amendment—An organic or inorganic material added to reduce bulk weight and increase air voids allowing for proper aeration.
2. Energy or fuel amendment—An organic material added to increase the quantity of biodegradable organics in the mixture and, thereby, increase the energy content of the mixture.

Amendments that have been used with dewatered sludge include sawdust, straw, peat, rice hulls, manure, refuse, tree and lawn trimmings, vermiculite, and a variety of other waste materials.

A *bulking agent* is a material, organic or inorganic, of sufficient size to provide structural support and maintain airspace within the composting matrix. "Bulking particle" may be a more descriptive term. Such particles form a three dimensional matrix of solid particles capable of self support by particle-to-particle contacts. Sludge cake can be viewed as occupying part of the void volume between particles. If the bulking agent is organic, an increase in the energy content of the mixture is a secondary benefit. Wood chips are commonly used as bulking agents although other materials, such as shredded tires, are sometimes used.

Procedures for estimating the quantity of bulking particles required for structural conditioning are discussed elsewhere.* This discussion focuses on use of amendments and recycled product for structural and energy conditioning.

Energy Conditioning

Heat released from organic decomposition is used to heat the incoming solids, water, and air. Heat is also required to satisfy the latent heat of evaporation. Of these, the latent heat usually represents about 70 to 80 percent of the total energy demand. The system energy demand can be estimated with sufficient accuracy by multiplying the pounds of water evaporated by a factor termed H_e. The latter is an estimate of the system energy demand ex-

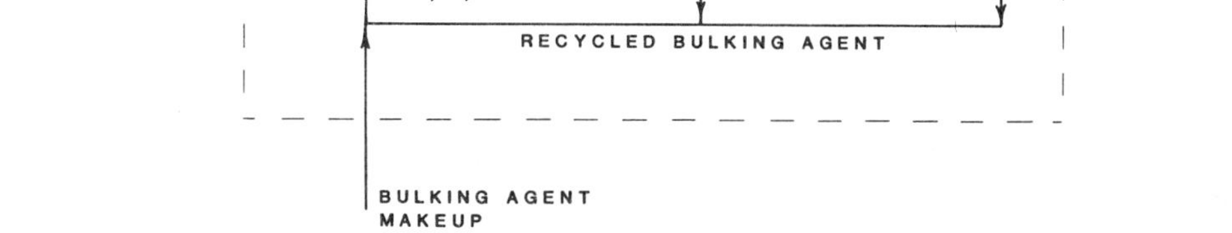

Fig. 1. Generalized mass balance diagram for sludge composting showing inputs of dewatered sludge cake, recycled compost product, amendment, and bulking agent.

pressed as Btu/lb of water evaporated. H_e is typically in the range of 1300 to 1500 Btu/lb water evaporated.

One approach to controlling the energy balance is to reduce the evaporation of water and, thereby, reduce the total heat demand. This can be accomplished by controlling and reducing the air supply to limit evaporation.

Control of the energy balance by limiting drying is a viable approach only if provision is made to adjust the final product moisture content by one or both of the following methods. First, moisture removal by air drying can be achieved in arid climates with open systems, such as the windrow process. In such cases, a dry product can sometimes be achieved, even though the starting sludge may not contain sufficient energy for complete moisture removal. Alternatively, heat drying can be used although this is not typically seen in practice. Second, a dry structural amendment can be added to produce an acceptable final product. Some composting systems are characterized by a low air supply rate, which reduces the evaporative burden and produces a relatively wet final product. Sawdust is often added in such cases. While the sawdust adds to the energy balance, it also serves to condition the final compost to an acceptable moisture content.

A second approach to controlling the energy balance—and one of the most effective—is to maximize cake solids achieved from dewatering. Sludge dewatering and composting are integrally related, and success in the latter depends heavily on the ability to achieve dry cake solids. Dewatering should be considered part of the feed conditioning process by both the designer and operator.

Increasing dewatered cake solids can improve the composting energy budget, however sewage sludges are difficult to dewater, particularly biological sludges from secondary treatment. In practice, it is often not possible to close the energy balance by dewatering improvements alone.

Another approach to controlling the energy balance is to add supplemental energy amendments, such as sawdust. This can improve the energy resources of the feed mixture to the point that these resources balance energy demands. When used in this way, energy amendment added to a composting mixture is analagous to supplemental fuel added to support a combustion process.

Role of Recycle: Recycle of compost product, or material from the first stage, does *not* add new BVS to the composting process. Material recycle is an internal loop within the system boundary and, therefore, does not add to the system energy balance. Any BVS remaining in the recycle were contributed by the original sludge or amendment. Since the latter are already included in the balance, including them again would cause a double accounting.

Structural Conditioning

The composting matrix is a network of solid particles containing voids and interstices of varying size. Voids between particles are filled with air and water. Free air space (FAS) is defined as the ratio of gas volume to total volume. The FAS of a composting mixture is important in determining the quantity and movement of air through the mixture. Available data indicate that FAS should be 20 to 35 percent to avoid rate limitations. The procedure for calculation of FAS for a mixture is presented elsewhere.[1]

Free air space in a sludge/amendment/recycle mixture becomes measurable at mixture solids contents above about 30 to 40 percent. In the absence of other data, a value of 40 percent is the recommended minimum solids content in the starting mixture for most sludge compost systems.

The following equation can be used to estimate the bulk weight of mixtures of sludge, amendment and recycle.

$$m = C/S_m \qquad S_m > 0.3 \text{ to } 0.35 \qquad (1)$$

where C = a constant in the range of 0.2 to 0.4. Values of 0.2 to 0.3 apply when lightweight amendments such as sawdust are a substantial part of the mixture. A value of about 0.4 applies when recycled material is the primary conditioning agent.

Conditioning with Product Recycle: Compost product can be recycled and used to condition the sludge cake. The quantity of recycled material required to adjust mixture moisture content can be determined from the process diagram of 1. Based on mass balance equations and assuming no bulking agent or amendment addition, the following solutions can be obtained for the daily weight of required recycle (X_r).

$$X_r = X_c\,(S_m - S_c)/(S_r - S_m) \qquad (2)$$

Let R_w be defined as the recycle ratio, based on total wet weight of recycle to total wet weight of dewatered sludge cake. Then:

$$R_w = X_r/X_c = (S_m - S_c)/(S_r - S_m) \qquad (3)$$

Let R_d be defined as the recycle ratio, based on dry weight of recycle to dry weight of dewatered sludge cake. Then:

$$R_d = (S_m/S_c - 1)/(1 - S_m/S_r) \qquad (4).$$

Conditioning with Amendments: Organic amendments can be added to a composting mixture for moisture control either with or without use of recycle material. Using the process diagram of Figure 1 and assuming no bulking agent or recycle addition, the following solutions can be derived for the daily weight of amendment required, X_a.

$$X_a = X_c\,(S_m - S_c)/(S_a - S_m) \qquad (5)$$

Again, let R_w be defined as the wet weight recycle ratio and R_d as the dry weight recycle ratio. Then:

$$R_w = X_a/X_c = (S_m - S_c)/(S_a - S_m) \qquad (6)$$

and

$$R_d = (S_m/S_c - 1)/(1 - S_m/S_a) \qquad (7)$$

The problem with using amendment for structural purposes is that amendment is usually more expensive than recycled product. Using amendment for moisture control without recycle can result in consumption of considerable quantities of amendment. Also, the quantity of final product is increased compared to systems which rely all or in part on compost recycle for moisture control.

Conditioning with Energy Amendments and Product Recycle

A compromise between use of product recycle alone (energy problem) and use of amendment alone (quantity problem) is possible. If recycle is used for moisture control, relatively smaller quantities of amendment can be added for control of the energy balance. In other words, amendment is added as the energy conditioner and recycle material added as necessary for further structural conditioning.

Again referring to the process diagram of Figure 1, there are three unknowns in the problem, X_r, X_a, and X_m. Three equations are required for solution and can be developed from balances on water, solids, and energy across the system boundary in Figure 1.

Water Balance:

$$\text{water in} = (X_c - S_cX_c) + (X_a - S_aX_a)$$
$$\text{water out} = (Xp - SpXp) = Xp\,(1 - Sp)$$
$$\text{water evap} = \text{water in} - \text{water out}$$
$$\text{water evap} = (X_c - S_cX_c) + (X_a - S_aX_a) - Xp\,(1 - Sp) \qquad (8)$$

Note that Equation 8 does not include the water of formation which is usually negligible.

Solids Balance:

$$\text{solids remaining} = \text{solids in} - \text{solids degraded}$$
$$SpXp = S_cX_c + S_aX_a - (k_cV_cS_cX_c) - (k_aV_aS_aX_a) \qquad (9)$$

Heat Balance:

$$\text{heat in} = \text{heat out}$$
$$(k_cV_cS_cX_c)\,H_c + (k_aV_aS_aX_a)\,H_a = (\text{water evap})\,(H_e) \qquad (10)$$

Substituting Equation 9 into 8 for Xp, substituting Equation 8 into 10, and rearranging gives the following equation for X_a:

$$X_a = \frac{H_e((X_c - S_cX_c) - (S_cX_c - k_cV_cS_cX_c)(1/Sp - 1)) - (k_cV_cS_cX_c)H_c}{(k_aV_aS_a)H_a + H_e((S_a/Sp - 1) - k_aV_a(S_a/Sp - S_a))} \qquad (11)$$

Once X_a is calculated from Equation 11, the mixture moisture content resulting from sludge and amendment alone can be determined as

$$S_m = (S_cX_c + S_aX_a)/(X_c + X_a) \qquad (12)$$

Let SMMIN = the minimum acceptable mixture solids content, usually about 0.40. Then,

if $S_m >= \text{SMMIN}$, then $X_r = 0$

if $S_m < \text{SMMIN}$, then recycle must be added.

The quantity of recycle to be added if $S_m < \text{SMMIN}$ can be determined as

$$\text{SMMIN} = (S_cX_c + S_aX_a + S_rX_r)/(X_c + X_a + X_r)$$

solving for X_r,

$$X_r = \frac{\text{SMMIN}\,(X_c + X_a) - S_cX_c - S_aX_a}{(S_r - \text{SMMIN})} \qquad (13).$$

Use of amendment for energy conditioning and product recycle for additional structural conditioning represents an optimal approach to adjusting the composition of the infeed mixture. Amendment use is reduced to the minimum required to satisfy the energy balance. FAS can then be adjusted as necessary using product recycle which is available at no cost to the operator.

References

1. Haug, R.T., "Compost Engineering—Principles and Practice", Library of Congress Cat. No. 79-56119, Technomics Publishing Co., Lancaster, PA (1980).

Detention Time

Attention has always been directed to detention time as a criterion for design and operation of composting systems. Perhaps this is a logical outgrowth of work on liquid phase systems, where detention time is a fundamental parameter of design. Unlike liquid phase systems, however, the interpretation of detention time and actual values required for design have not been universally accepted in the case of composting.

This section applies the concepts of hydraulic retention time (HRT) and solids residence time (SRT) to the composting process. Definitions and design criteria for each are proposed. Based on these criteria, the first-stage, high-rate and second-stage, curing phases of composting can be designed and integrated into a complete composting system.

HRT: HRT for a composting system can be defined as: The single-pass, mean residence time of the mixed materials, including recycle. [*Editor's Note:* Dr. Haug's complete discussion on detention time includes formulas for determining both HRT and SRT. See Sept., 1986 *BioCycle.*]

Design Criteria for HRT: Theoretical studies were undertaken to determine the effects of HRT on the composting process. Basic assumptions used to develop results presented here are as follows:

- Digested sludge is dewatered to 30 percent solids, with 50 percent volatile solids, and 50 percent degradability of the volatile fraction.
- Compost product at 70 percent solids is recycled to condition the feed sludge to a mixture solids content of 45 percent.
- Composting is conducted in a well-mixed, agitated solids bed reactor which is modelled as a continuous feed, complete mix process.
- No further BVS reduction occurs from the time material is output from the reactor until a portion is recycled.
- Air supply rate is constant during the composting cycle.
- The temperature of input sludge, recycled material, and air supply is constant at 20°C.

The effect of increasing reactor HRT is to increase the range of air supply over which adequate composting temperatures can be maintained. Both output solids content and BVS stabilization increase with increasing HRT for a given air supply. In other words, process stability is increased.

Thermophilic temperatures are predicted with detention times as low as two days. The advantages of a short detention time are, of course, the reduced reactor volume required and the associated capital cost savings. However, these savings are offset by the smaller range of operating parameters. Output solids content and BVS decomposition are both reduced at the shorter detention times. Clearly, the tradeoff is between capital cost savings and operational performance.

It should be noted that subsequent curing of the first-stage product would be required to produce a sufficiently stable material for most uses, particularly at the lower detention times. Therefore, capital cost savings from reducing the reactor HRT would be offset to some extent by increased curing requirements.

Reducing the infeed temperatures of solids and air to 0°C, simulating winter conditions, yields a similar trend of results. Longer detention times result in a greater range of operating air supply, higher output solids content and improved BVS stabilization. For a given detention time, however, the range of air supply for reactor temperatures > 60°C is reduced compared to the case with an infeed temperature of 20°C. At a 7-day detention time, for example, temperatures > 60°C are maintained at air supplies less than about 9 m^3/min-dry tonne/day (288 ft^3/min-dtpd) if the infeed temperatures of solids and air are 20°C. This is reduced to about 6 m^3/min-dry tonne/day (192 ft^3/min-dtpd) with infeed temperatures of 0°C. The effect appears to be more pronounced at shorter detention times, indicating an increased sensitivity to operating changes.

HRT is an important parameter that affects composting temperature, output solids content, and the stabilization of BVS. Process stability is favored by longer detention times. A minimum design HRT of 12-20 days is a reasonable choice for most reactor systems because the process is stable over a wide range of conditions and high output solids contents and BVS reductions are possible. Longer detention times produce smaller improvements in these parameters for a comparable increase in reactor volume. Many reactor systems are designed for an HRT of about 14 days which is consistent with the above discussion. Batch processes may require longer detention times because of the lag phase encountered at the start of composting.

SRT: SRT for a composting system can be defined as: The mean residence of the feed solids, excluding recycle.

Design Criteria for SRT: The minimum system SRT for design is likely a function of at least the following:

1. The extent of process control incorporated into the design and the processes used for the high-rate and curing phases,
2. The types of amendments used and whether the feed sludge is raw or digested,
3. The extent to which kinetic rate limitations can be avoided,
4. The end use of the product.

Time/temperature data were reviewed for a number of composting systems and substrates. Design parameters used for successfully operating systems were also reviewed. These data, together with the data referenced above, suggest that a minimum system SRT of about 60 days is required to produce a compost with sufficient stability and maturity to avoid reheat and phytotoxic effects. This assumes that the feed is properly conditioned with typical amendments and recycled product to close the energy balance and reduce kinetic limitations. It further assumes a reasonable level of process control to prevent excessive rate limitations from developing during the process.

An SRT of 60 days should be adequate for most composting systems and end product uses. However, longer SRT's may be required where a very stable end product is desired, for example, with high end uses such as in landscaping and horticulture.

Eventually, it may be possible to "fine tune" the minimum design SRT based on the process types used, characteristics of the feed substrates and amendments, and product end use. Until that time, a minimum system SRT of 60 days is recommended for design purposes.

Integrated System Design: A calculation procedure is presented which allows for an integrated approach to design of the first-stage, high-rate phase and the second-stage, curing phase. The procedure consists of the following principal steps:

1. Select the minimum system SRT to achieve the desired end product stability (SYSSRTMIN).
2. Select the desired HRT for the first-stage, high-rate phase (REACHRT).
3. Determine the mixture components necessary to achieve proper feed conditioning (see Part 1 of this series of papers).
4. Determine the volume of input and output materials to each stage of the process.
5. Determine the first-stage volume to achieve the desired HRT (REACVOL).
6. Determine the SRT for the first-stage process (REACSRT).
7. Determine the required SRT for the second-stage, curing phase (CURESRT), equal to system SRT minus first-stage SRT (SYSSRTMIN—REACSRT).
8. Determine the required second-stage, curing volume (CUREVOL) to maintain CURESRT.

Aeration

Once the feed sludge has been properly conditioned and the first-stage and second-stage composting processes have been sized, the next point of design control is the rate of aeration. Air must be supplied to a composting material for three basic purposes: (1) to satisfy oxygen demands imposed by organic decomposition, (2) to remove moisture from the composting material to provide drying, and (3) to remove heat generated by organic decomposition to control process temperatures. The ability to control aeration is one of the key points of process control and is an important consideration in process selection.

Air Demands: [*Editor's Note:* Formulas are provided to calculate stoichiometric demand and air demands for moisture and heat removal in his October, 1986 *BioCycle* report. The following is a summary of Dr. Haug's comments with regard to air demand. The discussion on aeration rates is presented in its entirety.]

- The moisture carrying capacity of air increases with increasing air temperature.
- Relative humidity of the inlet air is not a significant variable affecting drying if the exhaust gas is about 25°C or more greater than the inlet air temperature.
- The air requirements for drying and process temperature control are significantly greater than the requirement for biological oxidation. Either drying or temperature control can govern the air requirement depending on process conditions.
- At cake solids of 20 percent, the air requirement for drying can be 10 to 30 times that for biological oxidation. The different air requirements for biological oxidation and moisture removal allow control over the extent of drying by control of the air supply.

Aeration Rates

It should be noted that the weight of air moved through the compost process can be 30 to 50 times that of the dry weight of sludge. Composting has often been described as a problem of materials handling. The designer and operator should be aware that input air and exhaust gas are usually the largest components of the mass balance.

Batch Processes

For a batch process, such as the windrow and aerated pile, the total air *quantity* can be converted to a *rate* of aeration by considering the time duration of the process. Assume a batch process with a total air demand based on heat removal of 43.0 lb air/lb dry sludge. Assume that this quantity is supplied at a constant rate over a 25 day composting cycle. Then the *average* rate of aeration can be estimated as:

$$(43.0)\ (2000)/((25)\ (0.0748)\ (24)) = 1920 \text{ scfh/dt sludge}$$

where 0.0748 is the density of air at standard conditions, here defined as dry air at 20°C (68°F) and 760 mm Hg. (scfh/dt = standard cubic feet per hour/dry ton).

A number of factors can cause the *peak* rate of aeration to exceed the average rate calculated above. First, the rate of organic oxidation, and therefore, the rate of heat release, will vary throughout the composting cycle. Second, the type of aeration control may increase the peaking factor. An on-off aeration control logic is used in some systems. Obviously, the total air demand can be supplied only during the "on" sequence. Temperature feedback logic is popular with many composting systems and can also result in high peaking factors.

The peak rate of air supply can be estimated from data developed by Wiley,[1] Jeris, et al,[2] and Schultz.[3] These researchers observed that the rate of oxygen consumption is a function of temperature. Rates of 4 to 5 mg O_2/gm Vs-hr were observed in the temperature range of 50 to 65°C. For the feed mixture presented in Figure 2 this is equivalent to 0.0034 to 0.0043 lb O_2/lb dry solids-hr. The heat release rate can be estimated as:

$$(0.0034 \text{ to } 0.0043)\ (2000)\ (5866) =$$
$$39{,}900 \text{ to } 50{,}400 \text{ Btu per hr/dt feed solids}$$

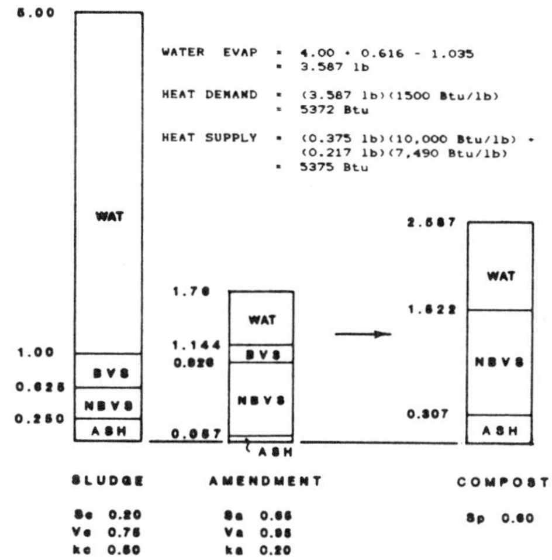

Fig. 2. Energy balance for the production of compost from dewatered sludge cake. Balance closed by addition of an energy amendment.

Using the same calculation procedure for heat removal presented above, the aeration rate required to maintain 55°C is:

$$x\ (113.5 + 3.90 + 19.4) = 39900 \text{ to } 50400$$
$$x = 292 \text{ to } 368 \text{ lb air}$$
$$\text{per hr/lb feed solids}$$

Again assuming an air density of 0.0748 lb/ft^3 at standard conditions, the peak aeration rate becomes:

$$(292 \text{ to } 368)/0.0748 =$$
$$3900 \text{ to } 4900 \text{ scfh/dt feed solids}$$

Using a theoretical model of the aerated pile process operated with temperature feedback control logic, Haug[4] predicted peak aeration rates of about 3800 to 4800 scfh/dt of sludge for raw sludge blended with wood chips. Murray, et al[5] reported that peak aeration rates of 4000 to 5000 cfh/dt were sometimes insufficient to keep process temperatures below 60°C when composting a raw sludge/wood chip mixture. Both the theoretical and measured values compare favorably with the estimate based on oxygen consumption data.

It should be noted that the peak aeration rate may be maintained for only a short period until the peak demand has passed. In the model presented by Haug[4] the required aeration rate exceeded 4000 scfh/dt for about 2 days, 3000 scfh/dt about 4 days, and 2000 scfh/dt about 8 days.

If the aeration system cannot meet the peak demand, process temperatures will exceed the desired set point. The designer must make a tradeoff between aeration system capacity, and hence capital cost, and the needs for process temperature control. It may be more cost-effective to size the system for less than the peak demand and accept process temperatures above set point for a short period. Dynamic process models are useful to develop the statistical information, such as that presented by Haug[4] above, on which to base such a decision.

Aeration systems for batch processes take many forms.

Some aerated pile and windrow systems use a single blower connected to a single pile. In this case, the peak demand is carried by the blower. Other systems use a central blower facility connected to many individual piles. In this case, the peaking factor can be "shaved" because it is unlikely that all piles simultaneously would be at the point of maximum oxygen consumption.

Minimum aeration rates should also be considered because the system must be capable of operating over the turndown range from peak to minimum rates. For an individual pile or windrow the minimum average rate usually occurs at the very beginning of the batch cycle and may be as low as 200 to 500 scfh/dt sludge.[6]

Continuous Processes

Many reactor systems operate on a continuous or semicontinuous feeding schedule. Typically, sludge is conditioned and loaded into the reactor on a daily basis. Reactor HRT is usually greater than 10 days, so that a daily feeding schedule can be modeled as a continuous process.

For a continuous feed process, the quantity of required air can be converted to an *average* rate of aeration by considering 1 lb/day of feed sludge. Using the air quantity for temperature control determined above, the aeration rate becomes:

43.0 lb/day of air per lb/day of dry sludge

This is equivalent to 47,900 scfh/dtpd of sludge (standard cubic feet per hour/dry ton per day).

Using dynamic process models, Haug[6] predicted maximum sustainable aeration rates between 40,000 and 60,000 scfh/dtpd of sludge for an autogenous raw sludge/recycle mixture composted in a continuous feed, complete mix reactor.

With continuous feeding, the *peak* aeration demand should not exceed the average demand over the composting cycle (i.e. peaking factor = 1). With semicontinuous feeding, some peaking might occur between feed periods. However, in practice it has not been necessary to consider this effect. The 15 to 20 percent contingency normally applied to blower design capacity appears to be sufficient to handle any above average aeration demands.

The designer should consider that feed quantity and composition may vary throughout the life of the project. The process should be examined under all expected conditions to determine whether seasonal or periodic peaking factors should be applied. Similarly, minimum flow conditions should be considered. This is particularly important if the process will operate below design loadings in early years. There is a tendency to focus attention only on average or peak design loadings. However, the full range of operation should be considered to avoid control problems during minimum turndown conditions.

Controlling The Aeration Rate

Once minimum, average, and peak aeration rates have been determined, the designer must select a control system to regulate the aeration rate. There are a number of control strategies which have been used in practice. These can be summarized as follows:

1. **Uncontrolled.** Windrow aeration by natural ventilation is an example of this approach. Such an approach is not applicable to aerated pile and reactor systems.
2. **Manual.** Throttling valves can be used to manually control the aeration rate. This is a common practice where lateral ducts are tied to a manifold which in turn is connected to a central blower system. Flow through each lateral is manually readjusted by a throttling valve throughout the composting cycle.
3. **On-Off Sequencing by Timer.** With a fixed blower capacity the average rate of air supply can be controlled by regulating the blower "on-time" by means of a timer control. This approach is often used with the static pile system and some reactor systems.
4. **Feedback Control Based On O_2 or CO_2 Content.** Oxygen or carbon dioxide probes inserted into the composting mixture can generate a control signal used to modulate air supply and maintain a setpoint O_2 or CO_2 content. O_2 or CO_2 is the *controlled* variable and aeration rate the *manipulated* variable. One problem with use of O_2 as a control signal is that the O_2 content often differs only slightly from that of ambient air because of the high air quantities required for moisture removal and temperature control.
5. **Feedback Control Based on Temperature.** Thermocouples inserted into the composting mixture can generate a control signal used to modulate air flow rate or blower "on-time" and maintain a set-point temperature. Temperature is the *controlled* variable and aeration rate the *manipulated* variable. This control approach is used in some aerated pile and many reactor systems and is considered well proven.
6. **Air Flow Rate Control.** If the daily quantity of feed material remains reasonably constant, it is possible for the operator to control air flow rate to a fixed value by either manual adjustment of fan dampers or by a setpoint flow rate control loop.

Where a control signal (temperature for example) is used for regulation, the air supply can be adjusted by a number of techniques. Throttling of inlet guide vanes or inlet dampers is commonly used with fixed speed aeration blowers. Outlet dampers can also be used. In the latter case, the possibility of fan surging at low flow conditions should be considered. A variety of variable speed drives can also be used to adjust the aeration rate. Speed control becomes more cost-effective in large systems where the power savings can offset the added capital cost.

It is important to consider minimum, average, and peak aeration rates to define the range of system turndown. Flow measuring devices and throttling valves must operate over the entire range of expected conditions.

References

1. Wiley, J.S., "A Preliminary Study of High-Rate Composting", Trans. ASCE, Paper No. 2895.
2. Jeris, J.S., Regan, R.W., "Controlling Environmental

Parameters for Optimum Composting,'' *Compost Science,* (Jan-Feb, 1973).
3. Schultz, K.L., ''Continuous Thermophilic Composting'', *Compost Science,* (Spring, 1962).
4. Haug, R.T., ''Modeling of Compost Process Dynamics,'' Proceedings of the Nat. Conf. on Composting of Municipal and Industrial Sludges, Wash. D.C., published by Hazardous Materials Control Research Institute, Silver Spring, Maryland, (May, 1982).
5. Murray, C.M., Thompson, J.L., ''Strategies for Aerated Pile Systems'', *BioCycle, Journal of Waste Recycling,* Vol. 27, No. 6, (July, 1986).
6. Haug, R.T., ''Compost Engineering—Principles and Practice'', Library of Congress Cat. No. 79-56119, Technomics Publishing Co., Lancaster, PA, (1980).

7

PHYSICAL, CHEMICAL AND BIOLOGICAL RELATIONSHIPS

Four members in the Department of Environmental Science at Rutgers University wrote a series of articles, "Analysis of EPA Guidance on Composting Sludge," in 1987 in BioCycle. *The researchers who authored the series are M.S. Finstein, F.C. Miller, J.A. Hogan and P.F. Strom. The topics which they analyzed included biological heat generation and temperature; biological process control; oxygen, moisture, odor and pathogens; and facility design and operation. Following are excerpts from that series:*

COMPOSTING is based on the phenomenon of spontaneous (self-initiating) biological self-heating, manifested by a temperature increase in a pile of organic material. The heat, generated through microbial metabolism at the expense of the organic material, accumulates in the material, increasing its temperature. The responsible microbes (various bacteria and fungi) are reliably present in almost all organic materials, and the physical and chemical requirements for initiating the process are not stringent. Hence informal, "backyard" composting of leaves and garden residue is both possible and popular, and at this level a casual approach to process control suffices. In contrast, large scale composting of wastewater sludge is highly demanding and should fully exploit the biological potential of the process.

Heat and Temperature

A crucial point in understanding composting process dynamics is the distinction between heat and temperature. A helpful reminder is that heat is measured in such units as calories or BTUs, and temperature in degrees Celsius (centigrade) or Fahrenheit. Heat is a form of energy. Its measurement is difficult. Temperature represents a potential driving force for the transfer of heat energy. Measurement of temperature is a simple matter.

Heat Removal

Heat removal and its effect on temperature play a crucial role in the behavior and management of the composting system. In large-scale systems ventilated by blower, the main avenue of heat removal is ventilation. Two ventilative mechanisms are operative: the vaporization of water (evaporative cooling) and sensible heating (temperature increase of air). Vaporization is dominant, accounting for approximately nine-tenths of the heat removed. (Assumptions: inlet air, 20°C and 50% relative humidity; outlet air, 60°C and 100%RH.) Conduction, a nonventilative mechanism, generally removes only minor amounts of heat from large deliberately ventilated systems.[1,2]

Ventilation plays other critical roles in addition to removing heat and water vapor. These are to supply O_2 for aerobic respiration and to remove the resultant CO_2. Most of the heat generated in the composting system results from aerobic respiration, whereas other forms of metabolism (fermentation, anaerobic respiration) play a minor role in this respect.

Air Function Ratio

An important relationship concerns the amount of air needed to remove heat relative to that needed to supply O_2, which we call the *air function ratio.* Assume that respiration consumes all the O_2 in a unit of air, resulting in the generation of one unit of heat; approximately nine units of air are needed to remove this amount of heat.[3] (Assumptions: heat generated per mass O_2 in the complete oxidation of organic matter to CO_2 and H_2O, 14,000 kiloJoules per kilogram O_2; inlet and outlet air, same assumptions as above.) The air function ratio varies slightly, depending mainly on inlet and outlet air temperature and humidity.

Heat Generation-Temperature Interaction

An unusual and crucial feature of the composting process is an interaction between the biological factor of metabolic heat generation and the physical factor of temperature. The activities of microbes, like those of all living organisms, are profoundly affected by temperature. Composting is unusual among microbial systems in that its temperature results not only from ambient conditions but also from self-generated heat. The self-induced temperature increase characteristic of composting is at first favorable to heat generation and other manifestations of microbial activity (positive feedback), but in the absence of deliberate heat removal (and assuming that O_2 is not severely limited) temperature generally becomes unfavorably high, suppressing heat generation (negative feedback). Hence, from the outset of composting an interaction between microbial heat generation and temperature comes into play. It would be hard to overstate the conceptual and practical importance of this interaction.

Rutgers Strategy

In the late 1970s field studies on sludge composting were conducted in Camden, New Jersey, by several groups at Rutgers University under common contract to EPA. One group's plan was to develop a process control strategy based explicitly on the heat generation-temperature interaction, with the operational objective of restricting the temperature ascent to a maximum of 60°C. Implementation was in batch configuration without mechanical agitation or enclosure, thereby preserving the structural and operational simplicity originally motivating the Beltsville method. A deliberate approach to biological process control was employed, however, as manifested in the arrangement for ventilation. This was to actuate the blower via temperature feedback control to match ventilative heat removal to its generation, poising temperature at the desired maximum. Only the forced pressure mode of ventilation was used, as it proved superior to vacuum induction. This approach came to be termed the *Rutgers strategy.*

The material composted was primary wastewater sludge mixed with wood chip bulking agent. A later study demonstrated that wood chips could be replaced with recycled compost.

The initial field tests involved 16 7-ton piles and one 20-ton pile. The results were reported to the EPA and subsidiary sponsors in 1980 and in published papers starting in 1980.[1,2] At this point the following were established:

- Temperature feedback control of a blower can effectively maintain an operational ceiling of $<60°C$ in batch composting of a primary sludge, without mechanical agitation. Typically the temperature sensor was positioned approximately 2 feet above the point of air inlet in a 7-foot high pile, with a set-point of 45°C (blower actuated at higher temperatures).
- Blower capacity must suffice to meet peak demand for heat removal. For the mixture of primary sludge (volatile solids content, 75%) and wood chips tested, peak demand was approximately 3,700 cubic feet per hour (cfh) per wet ton of the sludge-wood chip mixture. This is roughly equivalent to 30,000 cfh/dry ton sludge.
- Back pressure was typically 2 inches of water head, permitting the use of cost-effective radial blade blowers; forced-pressure ventilation proved superior to the vacuum-induced mode.
- In a typical trial the level of O_2 in the interstitial atmosphere was usually $>15\%$, the moisture content of the mixture decreased from an initial level of 75% to 23% in 15 days, and the odor quality improved rapidly judging from odor panel evaluations of samples.
- During this 15-day period total rainfall was 5.1 inches. Because of the strong drying tendency this caused only superficial wetting.

These observations formed the basis of a coherent theory of composting process dynamics which accounts for key interactions among heat generation, temperature, ventilation, and vaporization.

The theory predicts and explains various aspects of system behavior, such as the following: a sharp temperature increase that occurred when blower operation was interrupted during a period of high demand; an observed relationship between the assigned temperature set-point and demand for blower operation; the comparative strength of the drying tendency at different set-points; the dwindling of demand in response to a depletion of microbially available water and/or substrate; a revival of demand with rewetting. The centerpiece of both the process control strategy and the theory of system dynamics is the interaction between heat generation and temperature.

Moisture Content

Water affects and is affected by numerous aspects of composting, as in the following examples: Water is both required for and produced by microbial activity; an increase in water content decreases porosity and gas diffusivity (decreases O_2 penetration into micropores); water is removed via vaporization (evaporative cooling), as driven by microbial heat generation. The removal of water from sludge is a major treatment objective because a dry process residue is more easily screened, stored, transported, and disposed of. More fundamentally, provided that temperature feedback is employed, water removal and decomposition are linked via heat generation and vaporization. In this circumstance, the removal of water signifies the decomposition of sludge (main treatment objective).

Suggested Basis of Guidance with Respect to Sanitation

Conducting a sanitary composting operation and producing a sanitized compost are complementary objectives. In pursuing these objectives the elements of rational guidance are as follows:

- The primary stage facility should be designed and operated, without compromise, to maximize the rate of microbial activity. This involves the use of ventilative heat removal to establish a strictly observed 60°C *ceiling*.
- Depending on the overall sludge management plan (including the means of ultimate disposal/resource recovery), a secondary curing stage can be employed for further stabilization and assurance of disinfection.
- If formal assurance of disinfection is required, this is based on a temperature *floor,* as monitored during curing.
- Disinfection is considered to be assured when the curing material attains at least 60°C for one day or more or at least 55°C for four days or more.
- Special measures are needed to ensure adequate time-temperature exposure of edge material. Such measures include (but are not restricted to): formation of large or back-to-back curing piles; use of old cured material to insulate edges; scraping edge material off and putting it into the center of a new primary or curing pile.
- The distinction between the ceiling of 60°C (primary stage) and the floor of 60°C or 55°C (curing stage) is to be strictly observed. During the primary stage no deliberate attempt should be made to provide assurance of disinfection, as this would unnecessarily compromise various treatment objectives.

These suggested time-temperature regimens, easily attained during curing, correspond to more than a 25 log reduction of f2 bacteriophage.[4] Note the absence of conflict with both the federal regulation and other treatment objectives. In fact, the criterion suggested herein is stricter than the federal one (corresponds to only 15 log reduction).

References

1. Finstein, M.S. et al. Sludge Composting and Utilization: Rational Approach to Process Control. Final Report to USEPA, NJDEP. (1980) Available from NTIS, Accession No. PB82 13623.)
2. MacGregor, S.T. et al. Composting Process Control Based on Interaction Between Microbial Heat Output and Temperature. *Applied Environ. Microbiol. 41*:1321–1330. (1981)
3. Finstein, M.S. et al. Waste Treatment Composting as a Controlled System. p. 363-398. Microbial Degradations (Biotechnology, Vol. 8) 1986.
4. Burge, W.D. et al. Criteria for Achieving Pathogen Destruction During Composting. *J. Water Pollut. Control Fed.* 53:1683–1690 (1981).

8

MICROBIAL ACTIVITY IN COMPOSTING

Vicky L. McKinley, J. Robie Vestal and Atal E. Eralp

THE bacteria and fungi which make up the microbial communities in compost are the factories and the catalysts responsible for degrading the putrescible organics, resulting in the stabilization of material.

Composting is the result of the actions of a microbial community which converts easily-degradable organic matter to more stable, humified forms and to inorganic products (e.g. carbon dioxide, water, ammonia, nitrate, and methane), giving off heat as a metabolic waste product. The microorganisms also incorporate some of the organic matter into new cell material in order to grow and divide. The physical and chemical environment surrounding the compost microbes is constantly changing, primarily as a result of the accumulation of their own metabolic by-products, including heat. In turn, the conditions of the environment in which the microbes are in contact can greatly affect their ability to grow, to metabolize, and to survive.

This is a brief review of a study which was designed to derive a better understanding of the effects of various physical and chemical factors on the microbial communities active during the composting of sewage sludge. The microbial community was examined as a whole using accurate and relatively non-selective techniques for measuring both microbial activity and biomass. This section describes the relationships found between microbial activity and biomass and the moisture content, pH, organic content, total nitrogen, total carbon, C:N ratios, protein, and temperature of the compost. Also reviewed are experiments conducted to specifically address the role of temperature during composting, including a determination of an optimal temperature range for the compost microbiota.

Composting was accomplished using the agitated solids method in a production-scale composting bin. The compost piles were approximately 6 by 24 by 3 meters, and were aerated from below. The starting material consisted of primary raw sewage sludge (undigested and dewatered) from Akron, Ohio mixed with bulking agents (hardwood bark chips and sawdust) and previously-composted sewage sludge.

Moisture: An optimum moisture range of 50 to 70 percent moisture (30-50% dry weight) has been reported for composting with forced aeration. The moisture content of the samples taken during this study ranged from 31 to 60 percent, with most of the samples falling within the cited optimum range. Thus, moisture limitation did not play a large role in determining microbial activity during this experiment. No correlations between moisture and microbial biomass or activity were found. The moisture content of the compost did, however, decrease over time during the first run. During the second run, all of the samples were taken near the surface of the pile, where there is typically little change in moisture.

pH: Sewage sludge generally has a pH in the range of 5.5 to 6.5. The loss of organic acids or the accumulation of ammonia may cause the pH of sludge to increase during composting. Optimal pH values for composting reported have ranged from 6.0 to 7.5. During this study, the pH of the compost sampled increased significantly over time in both composting runs. The initial pH values ranged from 6.5 to 7.0, while the final pH values were in the range of 8.2 to 9.0. The increases in pH during this study seemed to occur more rapidly in the low-temperature areas of the compost piles (25° to 50°C) than in the high-temperature areas (65° to 80°C). This difference was most evident in samples taken before the piles were turned, at which time the material becomes somewhat mixed and the temperature zones temporarily disappear. Temperature ranges of 40° to 60°C generally yielded the highest and least variable pH values, although consistent statistical correlations between pH and temperature were not always found.

C:N Ratios: Total carbon, total nitrogen and C:N ratios were determined for the samples taken during the second run. None of these measurements correlated significantly with microbial biomass or activity, or with any of the other physical-chemical parameters which were measured.

Organic Matter: The organic content of the compost decreased over time in the compost sampled during the second run, but did not correlate significantly with any of the other measurements made during the run.

Protein: The protein content of the compost sampled during the second run increased over time (protein was not measured during the first run). The increase in protein was slower at high temperatures than at low temperatures.

Although one might expect that the proteins originally present in the sewage sludge would be broken down during composting, the increase in protein found during this study and others may be due to an increase in microbial biomass. However, during this study, the protein content did not correlate with biomass as determined by the quantification of lipid phosphate in the same samples. This fact may be due to the high turnover rates associated with lipid phosphate, making it a good indication of the viable biomass present. Proteins turn over more slowly, and hence may be an indication of living and some proportion of dead microbial biomass, and thus it may accumulate over time. This could also explain why lipid phosphate correlated negatively with temperature much better than did protein. Since the temperature gradients in the piles fluctuate somewhat, particularly during turning, the more slowly-degraded biomass constituents, such as protein, may not accurately reflect the effects of temperature on the microbial communities at any one time.

Biomass: Lipid phosphate levels were generally very high in the starting material, and then decreased somewhat as composting progressed. Biomass declined most sharply at the higher temperatures, decreasing significantly with increasing temperature during the first run. The high initial lipid phosphate levels could indicate either a high initial standing crop of biomass from the sewage sludge, or a slower rate of lipid phosphate turnover due to a relatively inactive population. The former is the more plausible explanation, since microbial activity was also relatively high initially.

Lipid phosphate biomass correlated well with microbial activity during the first run, but did not correlate with moisture, pH, organics, or protein in either run.

Microbial Activity: Microbial activity decreased dramatically when temperatures rose higher than 45° to 55°C, and fell to nearly zero when temperatures exceeded 60°C during both composting runs. Activity was found to negatively correlate with temperature (high significance level).

These same trends were also seen in all of the preliminary experiments performed during this study. The results were very similar regardless of which substrate ([^{14}C]acetate, [^{14}C]glucose or [^{14}C]glutamate) was used for the incorporation assay.

Of all the parameters measured during this study, temperature had the most dramatic and consistent effect on microbial activity. Within certain temperature ranges (e.g. 60° to 65°C), relatively small changes in temperature were found to cause changes in microbial incorporation rates of one or more orders of magnitude. Compost samples from the 35° to 50°C areas of the piles generally exhibited the greatest microbial activity.

The results of these experiments suggest that temperature is the dominant physical-chemical parameter controlling microbial activity and biomass during the composting of sewage sludge. Samples taken from high-temperature zones (> 55° to 60°C) were consistently found to have very low rates of microbial metabolism compared with samples taken from lower-temperature areas.

Temperature Effects On Microbial Communities

When temperatures exceed an organism's optimal growth temperature, the organism quickly comes under severe stress. Temperatures slightly below the optimum can generally be tolerated better than temperatures slightly above the optimum. Although thermophilic microorganisms are optimally adapted to grow at temperatures greater than 55° to 60°C, they have been found to sacrifice some metabolic efficiency in order to do so. They generally cannot grow or respire as efficiently as microorganisms adapted to moderate temperatures (mesophiles). Thus, although thermophilic organisms exist which can live at the highest possible composting temperatures, it is probably not desirable from a metabolic (and efficiency of decomposition) standpoint to run any such biological system under such extreme conditions unless absolutely necessary.

In order to make an objective evaluation of the metabolic status (or health) of any microbial community, one must consider all of the organisms present under conditions as close to those of their natural environment as possible. Thus, non-selective techniques (those not involving the culturing of individual strains) are preferred for the analysis of both microbial activity and biomass.

These studies have yielded several results which should be of use in developing more efficient composting systems. First, temperature is a critical composting parameter. Small variations in temperature can affect microbial activity and biomass in composting sewage sludge much more dramatically than small changes in moisture, pH, organics, or C:N ratios. Thus, temperature control during the composting process is highly desirable, if not mandatory, for an efficient and consistent process.

Secondly, optimum temperatures for composting were consistently found to lie below 55°C, regardless of the sampling temperature or the stage of composting. As batch composting progressed, adaptation to temperatures above ambient (but below 55°C) did occur, but thermophiles did not play a significant role. And finally, significantly greater rates of microbial activity were found when sewage sludge was composted with average temperatures not exceeding 58°C, compared with sewage sludge composted at overall temperatures reaching 85°C. Although composting temperatures must be allowed to increase to levels which assure adequate pathogen reduction (generally 55°C for three days, U.S Environmental Protection Agency 430/9-81-011), it is apparent from these studies and others that temperatures higher than this also incapacitate the microorganisms responsible for carrying out the composting process. Thus, precise temperature control is necessary to provide pathogen reduction while maintaining a healthy community of composting microbes.

Robie Vestal and Vicky McKinley—at the time this research on microbial activity was conducted—were with the Dept. of Biological Sciences, University of Cincinnati. Atal Eralp is with the Environmental Protection Agency.

9

EVALUATING THE IN-VESSEL COMPOSTING OPTION

John M. Walker, Nora Goldstein and Ben Chen

INCREASING numbers of decision makers are choosing the in-vessel composting technology when seeking a sludge management solution. Given this trend, there should be a systematic, objective method for evaluating the in-vessel composting systems in the marketplace. This chapter suggests methods that can be used in the evaluation process.

In-vessel options for sewage sludge composting may be defined as those which are under roof, mostly enclosed and easier to fit for "total" off gas collection and treatment. These in-vessel systems are generally more automated than the open windrow or static pile systems. The more open systems are usually constructed and turned with mobile equipment. The in-vessel systems either involve periodic mixing of the composting mass during the active composting phase—dynamic systems—or involve plug flow without mixing—static systems.[1,2]

When asked why they chose in-vessel composting over another composting method and what primary criteria were used for evaluating the systems on the market, engineers and treatment plant operators interviewed (about 15 to 20 in all) gave the following reasons:

1) Odor control;
2) Space constraints at the site;
3) Process and materials handling control;
4) Better public acceptance due to the aesthetics/enclosure;
5) Less manpower requirements;
6) More consistent product quality.

Choosing a vendor of in-vessel composting equipment is an extremely difficult process. There are more than a dozen different systems on the market, six or seven of which are actively marketed in the United States. Further complicating matters, noted one engineer who evaluated a number of proposals, is the fact that there is little interchangeability between the systems. The site work required for each can be very different, as are the physical structures.

The pace at which communities are choosing in-vessel technologies, combined with the lack of a systematic method of evaluating the ability of these systems to perform in key areas, pointed to the need for information that provides a basis for comparison. The evaluation methods described should be used only as guidance for analyzing the in-vessel composting option. The goal is to sufficiently stimulate the awareness of vendors, evaluators and operators so that a more detailed and thorough analytical approach will evolve and improved design and operation will result.

The first step of this analysis was to define the goal of a community in selecting a system. These goals include odor control, public acceptability, economy and compost quality.

Important considerations are to attain an odor-free, efficient operation. In general, all sludge composting systems ultimately can be operated to achieve pathogen kill and obtain a stable product. However, the odor and costs of operation are not so easy to control, and maintenance and upkeep could be a serious problem.

Meeting the goals requires an analysis of the following areas:

1. Citizen participation, which helps gain vital acceptance.
2. The sludge and bulking agent—what kind of sludge bulking agents will be available for going into the system and how much variability will there be in these ingredients and the sludge.
3. Mixing and porosity—examining the impact on air movement and moisture release.
4. The air handling system, which affects the requirements for proper biological activity, i.e. pathogen kill, and adequate drying of the material.
5. Temperature and pathogen control—optimums are not the same for stabilization, moisture removal and pathogen kill.
6. Retention time, which has an impact on completeness of stabilization, moisture removal, pathogen kill and size.
7. Curing—goals can include drying, pathogen kill, additional stabilization, and storage.
8. The mechanical features—reliability, skill level required for operation, and ease of repair.
9. The process features, e.g. determining the overall mass balance, then the energy and moisture balances, all of which determine the mix ratios and thus the sizing and cost of the facility.
10. The flexibility and back up system.
11. The credibility of the vendor, including financial stability, system experience and specific guarantees.
12. The practicality (reality) of the marketing or distribution program.
13. Vendor's data profile—certifying temperatures and moisture levels throughout the vessel.

Each of these 13 areas directly impacts the ability to have a successful in-vessel composting system, and most impact the ability to control odor, the primary concern voiced by everyone involved with a composting facility.

Key Areas For Consideration

Citizen Participation

A thorough, objective evaluation process must provide citizens, as well as those officials who will be involved in the decision-making process, with: 1) An opportunity for meaningful input into site and facility selection, and 2) A good education on the options available. It is very important to get citizens who will be affected by the facility involved at the earliest moment in the evaluation.[3]

They should be told about the critical features of each system under consideration, and given a realistic evaluation of the strong and weak attributes. When a decision is finally reached by the community, there is a better chance that citizens, particularly those living near the facility, will be more accepting of the method chosen for treating the sewage sludge. In addition, the system should have a better chance of operating well technically, instead of being under the constant threat of closure due to politics.

Sludge and Bulking Agent Properties

At least four factors need to be considered when analyzing the sludge that will be composted:

1) Volatiles content—does the sludge have enough energy to compost? The volatile solids content indicates the potential energy content (BTUs) available to heat the material. This parameter may be estimated by ignition in an oven at 550°C.[4] There are advantages and disadvantages to composting an anaerobically digested sludge versus an undigested sludge. A major advantage with a digested sludge is that there is less potential for odors during composting or curing; the disadvantages are that it takes longer to achieve desired temperatures and drive off moisture unless additional energy-rich bulking agent is utilized.

There is minimal knowledge on the availability and biodegradability of the volatile solids during the composting process. Even though a certain amount of volatiles might be measured, that will not ensure that the amount will be available. The availability is affected by such factors as system design and compost age.

2) Moisture content, which directly affects the energy needed to vaporize and subsequently remove water. It is important, therefore, to select dewatering equipment carefully.

3) Fineness, which affects the structure of the mix; and

4) Contaminants, such as metals or toxic organics, which limit the marketability of the finished material.

The properties of the bulking agent must also be analyzed, including the volatile content, moisture content, fineness and structure, and contaminants (although usually not a problem). In addition to providing extra energy, structure and porosity, bulking agents are useful as a possible diluent for contaminants and moisture and as an ad-

TABLE 1.

Criteria For Evaluation

(1) Full-Scale Operating Experience
(2) System Reliability
(3) Downtime for Maintenance
(4) System Flexibility
(5) Estimated Life-Cycle Costs
(6) Ability to Privatize
(7) Performance Guarantees
(8) Vendor Salesmanship

TABLE 2.

Community Goals For In-Vessel Sewage Sludge Composting Operations

(1) Odor Free
(2) Compost:
—Dry enough
—Stable enough
—Pathogen free
—Marketable
(3) Economical & Reliable
(4) Publicly Acceptable

ditive of other properties such as nutrients that are desirable and available in the final product.

One factor that will likely vary for both sludge and bulking agents is the moisture content. When the moisture content of the sludge or possibly even that of the bulking agent varies considerably from batch to batch, provisions must be made in the system for adjusting the ratio of sludge to bulking agent to gain proper composting. In some systems, adjustments (reductions) have had to be made in the amounts of recycle used relative to new bulking agent because of wetter than expected recycled material.

Likewise, adjustments must be made if sludge or bulking agent volatile contents vary. An example will be given in the mass balance section that shows how varying the solids content alters the amount of moisture that must be removed. And, as other parts of this discussion will indicate, there must be enough air, porosity, and matrix energy available to drive this moisture off.

Mixing and Porosity

The nature and extent of mixing have an important bearing on the porosity and the exposure of surfaces to aerobic breakdown and release of moisture. For example, very complete mixing of the initial sludge with recycle and new bulking agent materials (e.g. with a pug mill as compared to a front-end loader) will uniformly coat the sludge on the bulking agent[5,6] and in general, significantly decrease the energy required to force air through the system. (It is also important to be cautious about overmixing the materials.) Subsequent mixing, during the composting operation, exposes fresh surfaces and greatly accelerates the release of moisture compared with unmixed systems.[7]

The moisture level and fineness of the sludge, and the size of the bulking agent, also affect the porosity. The porosity and resistance to air flow are greater with a wetter and/or finer particle sludge and bulking agent and with greater distances for the air to travel.

Air Handling

Besides the shape and size of the reactors or bins, the aeration systems are probably one of the key differences between the in-vessel systems in the marketplace. Careful evaluation of this portion of an in-vessel system is extremely important because the handling of air—from beginning to end—can ultimately determine the success of a facility. It has a direct effect on odor generation, an operator's nemesis.

Air serves three main functions: 1) Moisture removal; 2) Assistance in stabilization (source of oxygen); and 3) Temperature control. When evaluating the air handling system, look at the source and the distribution. The air system should ultimately be compared on the bamix evenly to maintain aerobic conditions and control temperatures.

Air is supplied to different in-vessel systems in a variety of ways. Examples are given in 3. These systems should ultimately be compared on the basis of the amount of air required—at what horsepower and cost—to keep the temperatures at optimum levels. An example is shown in the mass balance section for sludges at several solid contents (e.g. 15, 20 and 25 percent). An initial mix solid content of some agreed upon value (e.g., 35 percent) must also be utilized in the calculations.

A calculation of the horsepower required to produce a given quantity of air for a given system can be useful. This horsepower requirement should be expressed on a basis that can be used to compare systems such as horsepower per cubic foot of air per dry ton of input sludge, as well as on a mix volume basis. However, this calculated value in and of itself will not indicate the suitability of the air flow, or the system, for achieving adequate moisture, temperature and odor control.

Air Collection

The collection of the air, after it has passed through the compost, should also be evaluated when looking at the air handling system. With air collection, three factors are important: 1) Configuration in the vessel; 2) Efficiency of moisture removal; and 3) Extent of condensation. A key factor, relating to thermodynamics, has an impact on air collection and handling: Air can hold four times more water vapor at 60°C than it can at 40°C. Therefore, if the temperature drops before the air leaves the vessel, it does not remove as much water and condensation within the vessel and composting mass will occur.

Some systems have an air collection system that extends into the composting mass. In some cases, parts of this collection system are often exposed to the air because of settlement of the composting sludge by natural compaction and microbial decomposition. Uneven subsidence may occur because of product removal that at times may be inconsistent, and because of friction along edge walls. When the air collection system is exposed, odors could escape and saturated air could cool with resultant condensate formation within the vessel.

Air Scrubbing and Odor Control

We can not overemphasize that the control of odor is one of the most critical elements of sludge composting. As stated earlier, pathogen kill and product stabilization nearly always can be obtained in any reasonable composting system. However, without odor control the operation will be shut down. There are a number of closed composting facilities that are monuments to this fact.

Odor should be controlled first of all by odor avoidance—proper aeration, control of temperatures, moisture level, porosity and mixing. These factors have already been described briefly in this paper and in more depth elsewhere.[8,9,10,11]

Second, all potential sources of odor should be identified and appropriately scrubbed. The scrubbing components of the air handling system need to be evaluated by looking at the exhaust air and gases at critical points. Malodorous air can be generated by numerous components of the system: Incoming sludge, sludge dewatering, the compost reactor, mixers, storage and hopper unloading areas, infeed/outfeed devices, and curing areas. Odors at composting facilities are caused by sulphur- and

TABLE 3.

Aeration Parameters In Various In-Vessel Systems

System	Type Air Source	Blower HP/CF of Compost Mass	Static Head Pressure, PSI	Straight Line Distance Air Must Travel Through Compost Mass*, FT	Volume of Air/CFM/CF of Compost Mass
A	Centrifugal or positive displacement compressor	0.002-0.008	1-3	21-25 plug flow	0.025-0.10
B	Axial flow fan	0.0006	Design 0.25, operate 0.14-0.18	8-11 dynamic	0.24 to 0.40
C	Positive displacement compressor	0.003	1-4	26 plug flow	0.09
D	Low pressure compressor	0.004-0.008	0.43-0.65	6-11 dynamic	0.24-0.40
E	Low pressure compressor	0.007	0.43	5-7 plug flow	0.085-0.12
F	Positive displacement compressor	0.002-0.008	2.5	1.5-27 plug flow	0.037-0.11

**Compost mass is not equivalent to amount and type of sludge composted and amount of moisture removed. This would have to be determined to finally compare these systems on a more equivalent basis.*
(Data in table has been verified by vendors 1986)

TABLE 3.

***Aeration Parameters in Various In-Vessel Systems** (Continued)*

	Air Diffusion Mechanism	*Air Collection and Scrubbing*	*Air Flow Direction*
A	Stone over pvc pipe grid; air to 1/4 of cylindrical system or all areas as required	Collected under vacuum at top of cylindrical vessel and blown through wastewater or scrubber as required	Up from most composted to least composted material
B	Stone over porous galvanized metal floor; air to all areas simultaneously as needed for temp. control (Rectangular)	Air not routinely scrubbed within vessel; provision to scrub up to 40% with reverse flow, 22% openings in deck	Up through equally aged compost mass; air can also be pulled down through
C	Stone over pvc pipe; aerated in 2 zones in each of several segments (Rectangular)	Negative pressure manifold about 3 ft. into compost mass at top; 1 to 1 outside air and process air into collection system; scrubbed as desired	Up from more composted to least composted material; Also pull air from fresh material at top downward
D	Stone over pvc or stainless pipe in concentric zones (Circular)	Positive or negative pressure; scrub as desired	Up through mixed
E	Stainless porous pipe over porous pipe; positive pressure	Stainless porous pipe over porous pipe; negative pressure; scrubbed as desired	Air flows from pipes—automatically alternated and periodically reversed between pressure and exhaust. Flow profiled to meet biological need
F	Diffusers across floor with fabricated orifices grouted into concrete floor	Same as diffusion; collect under negative pressure and scrubbed as desired	Air flow continuously in pressure and vacuum in each of 7 valved zones; relies on movement up into and then back down out of compost mass

nitrogen-containing compounds and other factors at the various process points.

How does the in-vessel system being evaluated treat all this air, and how effective is the system's method of treatment for controlling the types of odor involved? Vendors use various methods for scrubbing. These include compost/earth filters, chemical towers, aeration basins, and a general reliance on process control. The vendor and/or designer need to objectively describe the effectiveness of their collection and scrubbing systems for treating all potential sources and types of malodorous compounds. This is particularly important since not all vendors address all potential sources in their odor control techniques.

The location and size of the site for the in-vessel system is a third method of helping to deal with odor, but obviously this does not control the odor—rather it may make it less of a perceived problem. Smaller facilities usually have less material being composted that can potentially generate odor and more remote sites have fewer objecting neighbors. For facilities located where other sources of odor are present, e.g. near a sewage treatment plant, the identification and detection of any potential odor from the new composting facility is more difficult.

While enclosing a site helps gain public acceptance, it does not control the odor. Various in-vessel systems have also experienced varying degrees of problems with odor. In some cases, the problems have been lessened because of remoteness, being located at the treatment plant where there were other odors, or because of a more stable sludge being composted. In another case, an in-vessel system was temporarily closed because of the failure of the air collection and scrubbing mechanism. It is important to note that in the case of a more stable compost mix (containing a more stable anaerobically digested sludge), moisture may be more difficult to remove; and considerable problems have been encountered in the systems because of the build-up of moisture.

Temperature and Pathogen Control

Achieving certain temperatures at various points during the composing process is necessary to produce a stabilized, acceptable product. Listed here are the optimum temperatures needed to do the following.[5,6,7,8,9,10,11,12]

(a) Remove Moisture—40 to 55°C
(b) Stabilize—40 to 55°C
(c) Avoid Malodor—40 to 60°C
(d) Kill Pathogens—40 to 60°C (See EPA Criteria[12] for time/configuration requirements)

As is evident from this listing there are different optimum temperatures for different aspects of the composting operation. Optimum temperatures for (a) through (c) really are about 50°C or less and when maintained at this level, result in accelerated stabilization and removal of moisture with minimal production of odor. Optimum temperature for killing pathogens is higher. Make sure that sufficient energy is available in the mixture sometime during composting so that the EPA time/temperature requirements can be met.

Temperatures above 60°C (caused by low oxygen, high moisture level and/or poor porosity) should be avoided because they affect the survivability of the aerobic microorganisms that utilize the sludge as a substrate. It appears that above 60°C, the population of aerobic decomposing microorganisms decreases. Then, as the temperature decreases because of the reduced aerobic microbial decomposition, the aerobic population again begins to increase—and again, the temperature increases, the aerobic microbial population decreases and the cycling continues. Overall, therefore, the decomposition process is less effective and odors are generated by anaerobic zones that develop during the high temperature, low microbial growth part of the cycle.

Retention Time

Retention time within the composting reactor(s) has a direct effect on the size of the facility and the potential completeness of stabilization and moisture removal by the decomposing microorganisms. A variable that affects retention time, besides temperature, is how many times, on the average, the composted material (the recycle portion) has been through the system before removal.

The variables that determine how often the material can be recycled are its energy content and moisture level and the adequacy of air and mixing. Also important is the dryness and energy and moisture levels of new sludge and bulking agent used in the blend.

A balance is needed to ensure that the blend is always sufficiently dry and energy-rich to obtain a stable low moisture end product that is free of odor and pathogens in a reasonably sized facility. Such a balance can reduce costs both by holding down the size of the facility and the amount of bulking agent.

A formula has been described for calculating retention time, or "compost age".[13] Retention time can be conservatively calculated by dividing the total compost product discharged per day by the total product minus recycle (product for disposal) times the retention time for the system:

Retention time = Volume of Reactor(s)/ Volume of Mix per day

RT (days) = Vr/Vm

The "RT" calculated by this formula, however, is of limited value without knowledge of factors such as moisture levels, energy levels, and degree of mixing.

Curing

Curing is a vital factor that has an impact on retention time. Each system being marketed is designed around different amounts of time, and methods, for curing. Oftentimes, however, decisions about the curing stage, and whether it is done inside or outside the reactor, are very site specific.

Aeration and odor scrubbing also have to be considered during the curing phase (for both curing in the open or in a vessel). Goals for this phase should also be set, e.g. for further stabilization, for pathogen kill or assurance of adequate pathogen kill in all of the compost after remixing, for further drying and/or for storage. The evaluator should ask about the extent of mixing, aeration and/or air scrubbing to control odor during curing, and how likely the proposed system will be able to control odor and attain the other curing goals that were decided upon for their system's proposed input materials and design.

Mass and Energy Balance

A mass balance calculation is an important evaluation tool. It allows for systematically putting together many of the parts of the system to determine if what has been proposed can be achieved at what cost in what sized facility.

The mass balance establishes how much of each material—sludge, bulking agent or carbon source, and recycled compost—is used during each phase of the operation, and its fate during the process. This information is essential for determining design parameters such as the ratio of the sludge to bulking agent to recycle, the size of the equipment, storage needs, etc.

An EPA publication, "Composting of Municipal Wastewater Sludges",[14] provides a sample sheet for computing a mass balance. An adaptation of that sheet was prepared and computerized in EPA's Region IV Office.[15] We have included a partial presentation of data from this sheet (Table 4) to illustrate several of its important uses as an evaluation tool. Tables 5 & 6 offer examples of how a mass balance helps to determine if a system can actually do what you would want it to do (and what you are told it can do), e.g. remove "x" amount of moisture from the composting mass.

For example, assume that you, as an evaluator, have been told by a vendor that you have a fixed sized vessel for composting and that you can remove 4.8 pounds of water per pound of dry solids in this reactor (Column C, Table 5). Using the mass balance, you can test to see if the water removal rates make sense for various input mixes for a finished compost product of different moisture contents (Column E).

The test indicates that 5.4 lbs water/lb mix VS (Column D) must be removed to gain a compost product with 45 percent solids (Column E) when you use 50 percent sludge and 50 percent bulking agent (Column F) and the initial mix percent solids content is 34 percent (Column B). With the same beginning mix solids content of 34 percent, the amount of water that must be removed during curing jumps to 12 lbs/lb mix VS (Column D) when a 55 percent solids finished compost product (Column E) is required. It is very unlikely that you could get either of these amounts of water removed during curing if the curing vessel was the same as the composting vessel with respect to air flow rate, retention time and temperature—remembering that the maximum water removal during composting in that specific case in the example was 4.8 lbs/lb mix VS.

Reducing sludge solids (Table 6) from 23 to 17 percent (Column A) nearly doubles the required amount of moisture that must be removed from 32 to 51 lbs/lb mix VS (Column D) with a constant 50/20/30 ratio of sludge/bulking agent/recycle (Column F) in the initial mix. Based upon the previous reasoning discussed with regards to Table 5, the system could not remove the required amount of water without adjustments in critical parameters such as aeration and bulking agent.

We have extracted data (Table 7) from mass balance calculations by a vendor[16] to show how sludge and bulking agent solids contents have an impact on water removal, costs, and reactor size. The calculations show that reducing the sludge solids content from 25 percent to 15 percent (Column A) (with the bulking agent solids content remaining at 50 percent—Column B) would more than triple the cost for bulking agent (Column F) and nearly triple the required reactor capacity (Column G). The calculations show that air requirements (Columns H & I), the amount of water that would have to be removed

TABLE 4.

Mass Balance Sheet From (15) Computer Programs. Region IV, USEPA

	Compost Material Balance							
	Sludge	Bulking Agent	Recycle	Raw Mixture	Compost BioReact	To Curing	Compost Curing	Finish Compost
Volume Ratio: %...								
Bulk Density: lb/ft^3								
Solid content: % ...								
Total Solid: lbs (dw)/day........								
Volatile Solids: % lbs/day								
V.S.: lbs (dw)/day..								
Fixed Solids: lbs (dw)/day........								
Total Weight: lbs(wet)/day.....								
Water: lbs/day.....								
Moisture Content: %..............								
Volume: ft^3/day....								

TABLE 5.

Data from a Computer Analysis (15) Showing Required Moisture Removals to Obtain Different Percent Solid Compost Products

% Solids		Water Removed LBS/LB Mix VS		% Solids Finished Product	Sludge/ Bulk/ Recycle
Sludge A	Mix B	Composting* C	Curing D	E	F
20	34	4.8	5.4	45	50/50/0
20	34	4.8	12	55	50/50/0

**Assumes only 4.8 LBS of water/LB Mix Volatile Solids can be removed during the composting step.*

TABLE 6.

Data from a Computer Analysis (15) Showing How Varying Sludge Solids with a Constant Bulking Rate Impacts Required Moisture Removal

% Solids		Water Removed LBS/LB Mix VS		% Solids Finished Product	Sludge/ Bulk/ Recycle
Sludge A	Mix B	Composting* C	Curing D	E	F
23	41	4.8	32	55	50/20/30
20	37	4.8	41	55	50/20/30
17	33	4.8	51	55	50/20/30

**Assumes only 4.8 LBS of water/LB Mix Volatile Solids can be removed during the composting step.*

(Columns K & L), and the costs for power (Column M) would more than double.

The impact of a drier bulking agent (from 50 to 70 percent solid, Column B), keeping the sludge solids content at 25 percent (Column A), can also be seen on the various operating parameters and costs (Table 7). The savings in amendment cost (assuming the same cost/cubic yard) would be about 50 percent (Column F), and the power savings about 20 percent (Column M).

Other data extracted from the vendor's calculations (Table 8) show major impacts of air flow configuration on operations and costs. The assumptions here were that the same amount of air would be required for aerating a constant volume of biomass. However, the air configurations were different with the minimum air flow distance being five times greater in one of the systems. At a given sludge, bulking agent mix and product solids content (Columns B, C, D, & E) and with an increased air travel distance of four to five-fold (Column A), the calculations show a static head pressure increase of over ten-fold (Column G) and an energy requirement increase of about five-fold (Columns I & L) to remove equal amounts of water (Column H) with the reactor size constant (Column F) and air flow constant (Columns J & K).

Again, it can be clearly seen that drier sludge and bulking agent (25/70 vs. 15/50 percent solids, Columns B & C) have a major impact on water removal (over two-fold, Column H), reactor size (over three-fold, Column F), air flow (over three-fold, Columns J & K) and energy requirement (over three-fold, Columns I & L) for any given minimum air flow distance (Column A).

These calculations assume that there is no difference in the amount of water that condenses out of the air as it travels through the composting mass. If condensation were greater in one case than another, the amount of air required, costs, etc., would increase.

We have used data generated by others[16] in Tables 7 & 8. The point of our using these example mass balance calculations is to indicate the types of questions that can and should be addressed during the evaluation of potential in-vessel systems, and not to guarantee that the data or the conclusions reached about these example systems would be correct for your system. It is vitally important, when you use mass and energy balances to aid your evaluation,

TABLE 7.

Impact of Sludge and Bulking Agent Solids Contents (16) on:
A: Amount and Cost of Bulking Agent and Size of Reactor Required

% Solids				Sludge/ Sawdust/ Recycle	Sawdust Cost, $/DT Sludge	Reactor Size, 1000's CF
Sludge	Sawdust	Mix	Finished Product			
A	B	C	D	E	F	G
25	70	35	47	50/25/25	7.2	261
25	50	35	48	40/30/30	11.5	333
20	50	35	49	30/30/40	17.9	520
15	50	35	49	20/40/40	40.6	931

B: Amount of Air Required, Amount of Water That Must Be Removed, and Cost for Power

Sludge/ Sawdust, % Solids	Air For Water Removal CFM/LB Sludge VS/Day		Product Density, LBS/CF	Water Removal, LBS/LB Sludge VS		Power Cost, $/DT Sludge
	Inlet	Outlet		Composting	Curing	
A/B	H	I	J	K	L	M
25/70	0.44	0.55	42	2.2	1.6	4.5
25/50	0.56	0.70	38	2.5	1.8	5.4
20/50	0.88	1.09	37	3.7	2.7	7.8
15/50	1.57	1.96	32	5.8	4.2	13.4

Calculations based on assumptions:

- *Sludge 60% primary and 40% waste activated, 60% volatile fraction, 100,000 pounds dry solids/calendar day.*
- *Vessel retention time is 14 days composting and 14 days curing.*
- *Maximum straight line air flow distance is 7 ft., static head pressure is 12 inches of water, air temperature and relative humidity of inlet air is 48 degrees F and 70% respectively and air temperature and relative humidity of outlet air is 120 degrees F and 100% respectively.*
- *Power cost is $0.05/KWH.*
- *Amendment cost (sawdust cost) is $3.00/cu. yd.*
- *Minimum air for water removal.*

(The data in this table represent hypothetical case only. Evaluators need to use their own data when making these calculations.)

to carefully consider the validity of both the data and the assumptions that have been used. The tables provided can be used as a partial format for evaluating your own data. We also refer you to other sources[14,15,17,18,19,20] for learning more about the use of mass and energy balances.

Other Areas For Evaluation

Backup Alternatives and Flexibility

Any composting operation, be it in-vessel, aerated static pile or windrow, should have a reliable backup alternative. It enables the operator to reject a batch of sludge that may have a negative impact on the operation, and it is absolutely necessary in case of a breakdown in the system. Useful backup systems include land-spreading and landfilling.

Flexibility is important in terms of bulking agent, e.g. if a municipality plans to use sawdust, but then needs to switch to wood chips or another type of amendment. It is also important in the case of a system breakdown; operators should be able to easily unload the reactors or vessels to do repairs and have a backup to the materials handling system (e.g. dual process trains).

While flexibility should be a priority in the design of a system, too much may be wasteful in terms of the practicality of the operation and the overall capital costs. Be sure to have flexibility where it is necessary, however be realistic with regards to an operation's capabilities.

Vendor Evaluation

When selecting a vendor, consider the reliability of the components, vendor service and backing of the product, and the possibility of performance guarantees, e.g. for pathogen kill and/or odor-free operation.

The use of performance bonds is often mentioned as a

TABLE 8.

Impact of Straight Line Air Path and Solids Contents of Sludge and Bulking Agent (16) on: A: Supply Blower Pressure and Water Removal During Sludge Composting and Curing

	% Solids						Composting & Curing
Straight Line Air Path, FT	Sludge	Sawdust	Mix	Final Product	Reactor Size, 1000's CF	Supply Blower Pressure, PSI	Water Removal, LBS/LB Sludge VS
A	B	C	D	E	F	G	H
5-7	25	70	35	47	261	0.43	3.8
5-7	15	50	35	49	931	0.43	10.0
26-32	25	70	35	47	261	5.0	3.8
26-32	15	50	35	49	931	5.0	10.0

B: Blower Horsepower and Amounts of Air and Energy

Straight Line Air Path, FT	Sludge/ Sawdust, % Solids	Total Blower HP	Air For Water Removal CFM/LB Sludge VS/Day Inlet	Outlet	Electrical Energy/Yr. 1000's KW	1,000's $
A	B/C	I	J	K	L	M
5-7	25/70	178	0.44	0.55	1,163	58
5-7	15/50	638	1.57	1.96	4,142	207
26-32	25/70	916	0.44	0.55	5,986	299
26-32	15/50	3,264	1.57	1.96	21,323	1,066

Based on calculations assuming:

- *100,000 pounds/day dry solids sludge with 60% volatile solids (VS) content.*
- *Ambient air temperature and relative humidity of 48 degrees F and 70%; exhaust air temperature and relative humidity of 120 degrees F and 100%.*
- *Air moisture removal capacity is 0.08 pounds of water/pounds of air and the specific volume is 12.72 CF/lbs. air.*
- *Inlet air for composting and curing is: 8.5 CFM and 12 CFM per 100 CF, respectively.*
- *Minimum air for water removal.*
- *Electrical energy cost is $0.05/KWH*

(The data in this table represent a hypothetical case only. Evaluators need to use their own data when making these calculations.)

way to ensure that a facility will be built and will operate as desired. These bonds, however, are not a substitute for thoroughly evaluating the system being chosen; also, they can sometimes cloud efforts to do an objective analysis. It is still necessary, for example, to be sure the design does not violate the physical and biological principles of composting, that the specifications are reasonable with respect to energy availability, and that the systems have been tested to some degree. A community that accepts vendor information on blind faith and relies on performance bonds will be in a bad situation if the system does not operate as promised. These factors only make the need for evaluation methods more imperative.

Realistic Marketing

Compost marketing and distribution is not discussed here in detail, but it is important to mention as part of the evaluation process. The quality of the compost product is based on the following factors: uniformity, stability, pathogen-free, moisture content, contaminants, and lime. There are different applications for varying qualities of compost, e.g. compost used for landfill cover does not necessarily have to be of high quality, and a high lime compost is not good for acid loving plants.

Variables to consider when establishing a sales program are pricing (often compared to the going rate for top soil), customer service, who will run the marketing program—the public sector or the private sector—and whether the product will be sold in bulk and/or bags.

Operational Features

Other important considerations about the operational and mechanical features are listed below:

(a) Reliability of Components:
- —Infeed
- —Outfeed
- —Conveying
- —Screening

(b) Operators Required and Skill for Operating the System and Controlling the:
 —Mix
 —Air
 —Temperature
(c) Ease of Repair
 —Quick Removal of Compost Mix
 —Accessibility of Components
(d) Ease of Cleaning
(e) Freedom from Dust
(f) Automation
(g) Operate in All Weather (Cold & Wet)

While we will not elaborate further on most of these items since we have already discussed some of them in other sections of this paper as well as elsewhere[7,22,23], it is important to say a few words about the number and skill level of operators that are required. A generally assumed attribute of in-vessel systems is that the manpower requirement is reduced compared with open, less mechanical systems. This assumption should be carefully evaluated for each proposed system. The number of operators required for medium to larger sized in-vessel systems may indeed be less, but the skills required may be more.

For example, increased operator skill would be needed to properly use, maintain and calibrate the more sophisticated monitoring and controlling devices of in-vessel systems. Proper use requires sufficient understanding of these devices so that dial readings actually can be translated into appropriate blending and mixing; aeration, temperature and moisture control; stabilization of the mix; and odor prevention and removal. Higher level skills would also be required to repair and maintain the specialized mechanical equipment of the in-vessel system. Careful planning for recruitment, training and appropriate levels of pay are necessary to obtain and retain the necessary skilled personnel.

Moisture Content of the Mix when Discharged from the:
(a) Composting Phase
(b) Curing Phase (If separate)
Parameters:
(a) Air
 —Horsepower, Total/Specified Capacity of Sludge
 —Amount Air, CFM/Unit VS Sludge
 —Static Head Pressure, PSI
 —Configuration —Range of Straight Line Air Flow Distances
(b) Sludge, Bulking Agent & Mix Properties (for range of conditions)
 —Volatiles Content—Sludge, Bulking Agent, Mix
 —Initial Solids Content—Sludge, Bulking Agent, Mix
 —Porosity/Bulk Density—Mix, Product
 —Average Age of Mix
 —Type & Amounts of Sludge, Bulking Agent & Recycle
(c) Reactor Shape & Dimensions and Total Cubic Feet
(d) Mixing (When [Days] & How [During])
 —Composting
 —Curing
(e) Time (Days) In
 —Composting
 —Curing
(f) Mean Retention Time
 —Sludge
 —Bulking Agent

Fig. 2. Representation of Vendor's Operational Data.

References

1. Goldstein, N. 1985. Sewage Sludge Composting Facilities On The Rise. *BioCycle,* 26, #8:19-24.
2. Morales, L.M., J.C. Goldman, Jr. and J.C. Stowe, Jr., Oct. 1985. Selection and Design of In-Vessel Composting Systems—A Case Study. Presentation at 58th Annual Conference of the Water Pollution Control Federation in Kansas City, MO.
3. Yeaman, B., and J.M. Walker, 1985. Sludge Composting: Learning From Experience, Part III: Community Relations and Summary. *Operations Forum, 2,* #6:22-25.
4. Volatile Solids Determination, Method 208 G Standard Methods for the Examination of Water & Wastewater, 14th Ed. 1976. American Public Health Association, Washington, DC, pp. 96-97.
5. Kuter, G.A., H.A.J. Hoitink, and L.A. Rossman. 1985. Effects of Aeration and Temperature on Composting of Municipal Sludge in a Full-Scale Vessel System. *J. Water Pol. Cont. Fed.* 57:309-315.
6. Hoitink, H.A.J. 1985. Design Parameters for Composting at Small to Medium Size Treatment Plants. Paper Submitted to Ohio EPA from Ohio Ag. Res. and Dev. Center, Wooster, OH.
7. Yeaman, B., and J.M. Walker. 1985. Sludge Composting: Learning From Experience, Part I: Montgomery County's Site II. *Operations Forum 2,* #5:16-19.
8. Greeley & Hansen. January 1986. Odor Control Study of the Montgomery County Composting Facility for the Washington Suburban Sanitary Commission, Hyattsville, MD.
9. JTC Environmental Consultants, Inc. January 1984. Factors Affecting Odor Release During Sewage Sludge Composting for the Montgomery County Composting Facility for the Washington Suburban Sanitary Commission, Hyattsville, MD.
10. Miller, F.C., and M.S. Finstein. 1985. Materials Balance in the Composting of Wastewater Sludge as Affected by Process Control Strategy. *J. Water Pol. Cont. Fed.,* 57:122-127.
11. Sikora, L.J., and M.A. Sowers. 1985. Effect of Temperature Control on the Composting Process. *J. Environmental Qual.* 14:434-439.
12. USEPA. 1984. Environmental Regulations and Technology: Use and Disposal of Municipal Wastewater Sludge, Washington, DC, EPA/25/10-84-003.
13. Gossett, R.G. June 1985. Memograph on Compost Age. Compost Systems Company, Cincinnati, OH.
14. Alpert, J.E. 1985. Composting Facility Design. In Composting of Municipal Wastewater Sludges. EPA: 625/485-014.
15. Chen, B. Aug. 1985. Sludge Management Program and Sludge Applications/Composting System Design, Part Two. Presented at the USEPA Region IV Sludge Workshop, Atlanta, GA.
16. Lawrenson, J.G. Jr. Dec. 1985. American Bio Tech's Method of Computing Mass & Energy Balances, American Bio Tech, Jacksonville, FL.
17. Haug, R.T., and L.A. Haug. Nov./Dec. 1977 and

Jan./Feb. 1978. Sludge Composting: Engineering Principles Parts I & II. *Compost Science.* 18,6 & 19,1.

18. Haug, R.T. Aug. 1979. Principles of Sludge Composting. *Journal of WPCF.* 51,8,2189.
19. Haug, R.T. 1980. Compost Engineering Principles & Practice. Technomics Publishing Co., Lancaster, PA.
20. Haug, R.T. May 1982. Modeling of Composting Process Dynamics. Proceedings of National Conference on Composting of Municipal & Industrial Sludges, Washington, D.C.
21. Advertisement for Prequalification. 1985. Western Carolina Regional Sewer Authority, Greenville, South Carolina.
22. Yeaman, B., and J.M. Walker. 1985. Sludge Composting: Learning From Experience, Part II: A Problem Site, South Portland, Maine. *Operations Forum,* 2, #4:10-14.
23. Goldstein, N. 1985. Learning From Experience. *BioCycle,* 26, #1:22-25.3.

John Walker is with the Environmental Protection Agency in Washington, DC and is a member of the Editorial Board of BioCycle. *Ben Chen is with EPA Region IV based in Atlanta, Georgia.* BioCycle *executive editor Nora Goldstein coordinated the research.*

10

EVALUATING STATIC PILE AND WINDROW COMPOSTING

SEVERAL reports issued by the U.S. Environmental Protection Agency provide a good analysis of sludge composting. Two in particular include technological and economic evaluations of static pile and windrow composting.

Composting Municipal Sludge: A Technology Evaluation was issued in 1987 by the EPA's Water Engineering Research Laboratory's Office of Research and Development in Cincinnati, Ohio. It was written by Arthur Benedict, Eliot Epstein and Joel Alpert.

The report analyzes five municipal sludge composting facilities: the Hampton Roads, Sanitation District Peninsula Composting Facility located in Newport News, Virginia; the Washington Suburban Sanitary Commission Montgomery County Composting Facility (Site II) located in Silver Spring, Maryland; the City of Columbus, Ohio, Southwesterly Composting Facility; the Joint Water Pollution Control Plant Composting Facility of the Los Angeles County Sanitation Districts located in Carson, California; and the Metropolitan Denver Sewage Disposal District Number One Demonstration Composting Facility. The first three facilities listed above use the aerated static pile composting process. Los Angeles County uses a conventional windrow process and Denver's pilot (10 dry ton/day) was designed as an aerated windrow facility (The Denver plant now operates as a full-scale facility). No evaluation was done of facilities employing the in-vessel, mechanical composting technologies.

Sludge characteristics at the facilities studied are as follows:

- Hampton Roads: Anaerobically digested primary and secondary
- Site II: Limed, raw primary and secondary
- Columbus: Unlimed, raw primary and secondary
- Los Angeles: Anaerobically digested primary and secondary
- Denver: Anaerobically digested primary and secondary

The technology evaluation was performed to assess a broad range of municipal sludge composting activities related to design, operation, performance and economics. The authors therefore emphasize that the report is "intended to be used as an information source on municipal sludge composting rather than as a design manual." The scope of the evaluation ranged from on-site investigations consisting of observations of operations to a review of published information on the facilities. Some independent process testing was conducted as well.

The objectives of the evaluation were to compare and contrast features of the aerated static pile and windrow technologies based on experiences at operating facilities; assess operating, performance and cost features; identify key problems associated with municipal sludge composting using these technologies; and to define methods which have been used or are being considered to resolve these problems. The period covered by the report ran from January 1984 to October 1985.

The operations assessment for static pile composting

covers six major categories related to design and operation: 1) sludge loadings and characteristics; 2) sludge transport; 3) process technology; 4) physical design; 5) environmental controls; and 6) product disposition. The authors state that process technology features were of particular interest. These included bulking agent and/or amendment use, mixing methods and effectiveness, pile configuration characteristics, aeration requirements and temperature and oxygen monitoring and control requirements. Physical design features investigated include: 1) effectiveness of site layout from an operating perspective; 2) general materials management methods; 3) staffing, and 4) maintenance considerations. Actual capital and O&M cost information was obtained at each facility from which unit costs were developed. Similar details were assessed at the windrow composting facilities.

General Conclusions

Chapter 2 of the report presents conclusions of the municipal sludge composting technology evaluation. The general conclusions, based on a comparison of static pile and windrow composting technologies, are excerpted below:

Moisture Control—

Moisture control and a homogeneous initial mix were found to be the most important factors for effective composting. During day-to-day operations moisture must be controlled for effective stabilization, pathogen inactivation, odor control, and finished compost quality control. The moisture can come from several sources including: the incoming sludge, bulking agent, or amendment; inclement weather; and the composting process itself as a by-product.

An initial mix moisture content of 60% or less (total solids >=40%) was found to be a key criterion regardless of the composting technology employed. To be a marketable product the final compost must generally have 50 to 60 percent total solids.

Sludge Loading Variability—

Sludge loadings design must be carefully established and must consider the high variability of both the quantity and quality of sludge to be processed. For example, peak-to-average ratios for loadings at the static pile facilities studied ranged rom 1.4 to 1.9, and the total solids content of incoming sludge at these facilities ranged from 14% to 22%. Higher than anticipated values for moisture content, average daily loading, or peak day loading can cause several operational problems. Bulking agent or amendment use, mixing operations, personnel utilization, equipment utilization and effectiveness, area requirements, and composting performance can all be adversely impacted by loading variations. Therefore, operational flexibility to respond to the day-to-day variability in sludge total solids (moisture) content and the wet quantity of sludge received must be considered during design.

Although loadings based on calendar days are useful for general comparisons, operating-day loadings (for example, based on 5-day-, 6-day-, or 7-day-per-week operation) more accurately reflect day-to-day materials management and process requirements, and should be used for design purposes.

Wastewater treatment plant operations can also impact the effectiveness of composting operations. Variability in dewatered sludge total solids caused by changes in feed sludge characteristics to the dewatering operation is one example. Thus, integrating treatment plant and composting operations is important for effective composting performance. Timing sludge deliveries, coordinating plant operations with composting constraints, and providing sludge storage are examples of ways in which this can be accomplished.

Operational Area—

Unit operational area for static pile and windrow composting, expressed as acres per wet ton per day (ac/wtpd), is site-specific. Operational area is defined as site area required for process operations and support facilities such as administrative and maintenance buildings, but excluding extraneous areas such as land held on site for expansion purposes. Based on the facilities investigated, a reasonable range is 0.08 to 0.14 ac/wtpd, depending on such site-specific considerations as daily sludge loadings, storage requirements, drying technique employed, composting method, and the need for runoff control ponds.

Paving and Covering—

All of the facilities investigated have paved areas for most of their operations. Operating reliability is enhanced by the use of paved surfaces, based on wet-weather problems which have been experienced at the facilities studied. Covering or enclosing key operating areas can also enhance reliability; for example, all of the static pile facilities have protected mixing areas. However, site-specific and technology-specific factors are important in defining the need for protection at any given facility.

Equipment and Manpower Utilization—

Front-end loaders are used for a number of static pile and windrow operations, including rough mixing, pile construction, windrow formation, teardown activities, and materials storage or transfer operations. Mobile composters are used for fine mixing during static pile composting as well as for windrow turning.

Generally, manpower utilization is site-specific. However, based on the number of operating personnel employed at the facilities studied, the static pile technology is more labor-intensive than the conventional windrow technology.

Odor Control—

Odor generation and release is the main environmental problem at static pile and windrow composting installations, and odor management is a complex problem. A certain amount of odor is unavoidable, and a good public relations campaign should be used to minimize public opposition. However, some design and operational features can be employed to minimize the potential for generation or release of objectionable odors, including the following:

1. Trucks used to haul dewatered sludge should be covered and cleaned frequently. This is particularly important with raw sludge.

2. Dewatered sludge deliveries should be managed such that mixing and other operations can be performed without sludge accumulating for long periods of time. This is particularly important with raw sludge and in hot weather.
3. An initial mix moisture content of 60 percent or less is a key criterion for odor control. Enclosing operations and scrubbing exhaust gas can be an effective step for odor control in some situations.
4. A uniformly mixed and sufficiently porous material is important for odor control during composting. The presence of clumps of unmixed sludge can lead to anaerobic and, thus, odorous conditions, as well as incomplete stabilization and pathogen inactivation. Construction of daily static pile compartments, with cover material, is also important.
5. Simply limiting the daily quantity of sludge composted at a facility has been used as an odor control technique.
6. Positive aeration during active static pile composting utilizes the pile cover material as an odor scrubber. Negative aeration requires the use of a separate exhaust scrubber system such as a finished compost filter pile. Regardless of the aeration mode employed, aeration rate, oxygen content, and temperature must be controlled to promote active aerobic composting.
7. During the life of a conventional windrow, continuous emissions from the windrow surface are a greater source of odor than the periodic emissions during windrow turning. Although emissions are most intense immediately after turning, they then decrease rapidly. Surface emissions decrease as the composting process progresses.
8. Teardown of static piles or windrows can be managed to minimize release of odors. For example, teardown can be restricted during wet weather or during early morning hours when air inversions may occur. At static pile installations, high-rate aeration for 24 to 48 hours prior to teardown, which lowers pile temperature to ambient, can be used to control odor release.
9. Leachate, condensate, and runoff should be collected and disposed to minimize odor generation potential. Proper site drainage is required to prevent ponding which can generate odor.
10. Effective housekeeping procedures such as washing equipment and flushing or sweeping working areas also reduce odor generation potential.

Performance—

Both static pile and windrow composting meet pathogen inactivation and volatile solids reduction requirements and produce a marketable product while operating under a wide range of conditions. Sustained finished compost production between 0.4 and 1 cubic yard per wet ton (cu yd/wt) of sludge composted is representative of the facilities investigated. This production rate is equivalent to about 0.7 to 2 dry tons of compost per dry ton of sludge. (The apparent increase in mass is due to the incorporation of wood chip fines or amendment material into the finished compost.)

Facility Costs—

Representative costs in 1985 dollars for static pile composting are \$36 to \$51 per wet ton of sludge based on the facilities studied. These costs are higher than the costs estimated at design. Amortized capital costs based on a 20-year amortization period and an interest rate of 10 percent, and annual operation and maintenance costs, are included in these figures. Dewatering and sludge transport costs are excluded. Comparative costs for windrow technologies were not established during the technology evaluation. Operating cost reductions equivalent to \$1 to \$3 per wet ton are currently achieved from compost sales revenues at the facilities studied and all have marketing programs to increase revenues in the future.

Static Pile Conclusions

The technology evaluation also reached some conclusions specific to static pile facilities. These are listed in Chapter 2, and excerpted below:

Wood Chip Usage—

Dewatered sludge total solids content being received at each static pile facility studied is below that estimated at design by 3 to 5 percent. Because of this, wood chip use is as much as 80 percent over the design estimates. Increased wood chip use has increased composting and storage area requirements, constricted equipment mobility, and increased operating cost. Typical mix ratios are between 3.5 and 4.5 cu yd/wt, depending on season, proportion of new and recycled chips, and chip moisture content.

Initial Mixing—

In the absence of a standard test method, the use of a single person to oversee mixing operations ensures consistent and proper initial mix moisture content and mix homogeneity. Having more than one person responsible for mixing adversely impacts the effectiveness of the mixing operation and therefore should be avoided. A mixing time of 40-45 minutes is typically required.

Pile Construction—

Extensive testing following initial construction was required at two of the static pile facilities to arrive at reliable pile construction and aeration piping configurations for routine operation. Both uniform and variable hole spacing in the aeration piping is used. Extended static pile heights are typically 12 to 13 ft, but lengths and widths of daily pile compartments vary based on daily loadings. Minimum depths of base and cover materials are 12 and 18 inches, respectively. Two of the facilities use denser material (unscreened compost or dry, recycled compost) at the pile toes to prevent short-circuiting of air flow.

Composting Period—

A minimum active composting period of 21 days is required for effective composting at static pile facilities, but longer periods are sometimes used because of materials management considerations. Regulations require that internal temperatures at all monitoring points be maintained at $\geq$55 degrees C for at least 3 days for pathogen inactivation and degradation of putrescible organics.

Process Control—
Variation of aeration rates to achieve desired temperature and oxygen levels during active composting is the key process control method. Aeration rates per dry ton, number of blowers per pile compartment, individual blower capacity, total blower capacity on site, and blower capacity per wtpd vary from site to site. However, the trend is now toward providing one blower per daily pile compartment. Generally speaking, aeration rate is controlled by cycling blower operation until the time-temperature regulations are met. After this, high-rate, continuous aeration is used to lower the temperature and promote drying. Manual and automatic blower control are used, and both positive and negative aeration are used.

Drying—
Material composted by the aerated static pile method is usually not dry enough to screen directly, and a separate drying step is employed. A minimum unscreened compost solids content of 50 to 55 percent is generally required for effective screening. All of the static pile facilities investigated use some type of separate drying step and also use high-rate, continuous aeration during the final stage of active composting to promote drying.

Screening—
Screening effectiveness, including expected variability, needs to be carefully assessed in aerated static pile applications. Under optimum conditions, wood chip recoveries equivalent to 80 to 90 percent (based on unscreened compost volume) can be obtained. However, this requires that unscreened compost total solids content be between 50 and 60 percent, depending, in part, on screen size and design. Exceeding the upper limit can result in losses from excessive dust generation, while not meeting the lower limit inhibits the separation of fines from wood chips and leads to wood chip contamination, reduced screening rates, and often mechanical problems with the screening equipment. Wood chip recoveries of 65 to 85 percent are more typical of sustained routine operation at the static pile facilities investigated.

Curing—
Curing operations at static pile facilities are dependent, in large part, on materials management requirements. A minimum curing period of 30 days is provided, but longer periods are often used. Unscreened or screened compost curing can be used, but an advantage of screened compost curing is that it requires less area than unscreened compost curing. Aerated curing can also be effective.

Performance—
Finished compost having a total solids content of 51 to 69 percent and bulk densities of 900 to 1,500 pounds per cubic yard, and free of salmonella is routinely produced for sale at static pile facilities. Other characteristics, particularly metal content, are site-specific.

Operating Cost—
On-site labor, wood chip purchase, and aeration pipe replacement account for about 50 to 80 percent of the annual operation and maintenance (O&M) cost at static pile facilities. The annual O&M cost in 1985 dollars ranged from $25 to $36 per wet ton of sludge at the facilities studied.

Windrow Conclusions

Conclusions specific to windrow facilities studied are presented in Chapter 2 as well. They are as follows:

Windrow Formation: Initial rough mixing of sludge and amendment prior to fine mixing and windrow shaping using mobile composters is required for effective windrow composting. Equipment used for this purpose varies.

Composting Period—
A minimum active composting period of 30 days is required for windrow composting, but longer periods are sometimes used. Regulations require that internal temperatures at all monitoring points be maintained at ≥55 degrees C for at least 115 days. Use of large windrows with low surface-to-volume ratios (7 ft high and 23 ft wide at the base) during the latter stage of composting can conserve heat and aid temperature development for pathogen inactivation and degradation of putrescible organics. A turning frequency of two to three times per week is adequate for mixing and aeration. An additional curing step is not required.

Drying—
Based on an independent analysis of demonstration study results from the aerated windrow facility, induced aeration can improve drying by about 3 percent over conventional windrowing. The type of amendments used can also aid drying in some instances.

Dust Generation—
Dust generated from mixing, windrow turning, and sometimes windrow surfaces can be a problem.

Performance—
Windrow composting can routinely produce finished compost having a total solids content of up to 60 percent and free of salmonella.

Operation Cost—
Labor cost is the main operation and maintenance cost at windrow compost facilities.

Studies of odor control at the windrow operation in Los Angeles found that typically 83 percent of windrow odor emissions are the result of ambient surface emissions and 17 percent are the result of windrow turnings. These emissions decrease significantly as the active compost cycle progresses. Experience at the Los Angeles facility, say the report's authors, has indicated that the best way to control odors and minimize complaints is to limit the size of the composting operation. The maximum amount of sludge that can be composted during the summer without odor complaints is 500 wet tons per day. The type of amendment used in the windrow can also help to control ambient surface emissions.

Cost Evaluation

Details on capital and operating and maintenance costs can be difficult to obtain, however the authors of the EPA report were successful for the most part, especially for the static pile facilities. Total capital costs for the Hampton Roads operation (all numbers are presented in 1985 dollars) were about $2.6 million; for Columbus, total capital costs were about $6.6 million and for Site II, about $17.6 million. The authors had difficulty getting numbers for

Assumptions Used in Obtaining Costs and Requirements for Windrow Composting

Parameter	Assumed Value
Percent sludge solids in dewatered sludge	20 percent
Percent volatile solids in dewatered sludge solids	35 percent
Percent volatile solids destroyed during composting	30 percent
Percent solids in compost product	65 percent
Dewatered sludge specific weight	1,820 lb/yd^3
Compost product specific weight	865 lb/yd^3
Mixed dewatered sludge and compost specific weight	1,685 lb/yd^3
Windrow cross section	35 ft^2
Windrow length	300 ft
Truck unloading and mixing area	300 ft^2/ton/day dry solids
Finished compost storage area	900 ft^2/ton/day dry solids
Fraction of site requiring clearing (brush and trees)	0.7
Fraction of site requiring light grading	0.3
Fraction of site requiring medium grading	0.4
Fraction of site requiring extensive grading	0.3
Cost of site clearing (brush and trees)	$1,560/acre
Cost of light grading	$1,040/acre
Cost of medium grading	$2,600/acre
Cost of extensive grading	$5,200/acre
Cost of land	$3,120/acre
Cost of diesel fuel	$1.35/gal
Cost of labor	$13.50/hr
Cost of paving	$60,320/acre

the Los Angeles County facility, and Denver was operating in a pilot stage.

Annual O&M costs in 1985 dollars for on-site operations and administration at Hampton Roads, Columbus, and Site II static pile facilities are approximately $467,000, $1,048,000, and $3,483,000, respectively. The annual on-site O&M cost at the Los Angeles facility in 1985 dollars is $988,000. Information compiled by the authors shows that costs for on-site labor, including fringe benefits, comprise about 24 to 50 percent of the annual static pile O&M costs at these facilities, excluding sludge transport costs. Costs for bulking agent and aeration piping vary from about 20 to 40 percent of the annual O&M costs, excluding sludge transport.

The Los Angeles facility uses external amendments as well as recycled compost for conventional windrow composting, and there is no purchase cost for these amendments. Labor cost, at $624,000, say the authors, is the major O&M cost (63 percent), with fuel and equipment maintenance being 15 and 16 percent, respectively.

Assumptions Used in Obtaining Costs and Requirements for Aerated Static Pile Composting

Parameter	Assumed Value
Percent sludge solids in dewatered sludge	20 percent
Percent volatile solids in dewatered sludge solids	35 percent
Percent volatile solids destroyed during composting	45 percent
Percent solids in compost product	65 percent
Compost product specific weight	1,000 lb/yd^3
Mixed dewatered sludge and bulking agent specific weight	1,100 lb/yd^3
Bulking agent mixing ratio	2.5 yd^3/ton dewatered sludge
New bulking agent mixing ratio	0.625 yd^3/ton dewatered sludge
New bulking agent specific weight	500 lb/yd^3 dewatered sludge
Recycled bulking agent mixing ratio	1.875 yd^3/ton dewatered sludge
Recycled bulking agent specific weight	600 lb/yd^3
Truck unloading and mixing area	300 ft^2/ton/day dry solids
Composting area	7,000 ft^2/ton/day dry solids
Drying area	3,000 ft^2/ton/day dry solids
Finished compost storage area	900 ft^2/ton/day dry solids
Bulking agent storage area	2,000 ft^2/ton/day dry solids
Fraction of site requiring clearing	0.07
Fraction of site requiring light grading	0.3
Fraction of site requiring medium grading	0.4
Fraction of site requiring extensive grading	0.3
Cost of site clearing	$1,560/acre
Cost of light grading	$1,040/acre
Cost of medium grading	$2,600/acre
Cost of extensive grading	$5,200/acre
Cost of land	$3,120/acre
Cost of diesel fuel	$1.35/gal
Cost of electricity	$0.094/kwhr
Cost of labor	$13.50/hr
Cost of paving	$3.15/$ft^2$

The report presents unit O&M costs for each facility (excluding sludge transport). Hampton Roads has an annual sludge quantity of 14,500 wet tons or 2,465 dry tons at 17% total solids. Unit cost is $32 per wet ton in 1985 dollars or $189 per dry ton. Columbus' annual sludge quantity is 42,097 wet tons or 7,157 dry tons at 17% solids. Its unit O&M cost is $25 per wet ton, or $146 per dry ton. Site II's annual sludge quantity is 97,000 wet tons or 16,990 dry tons at 17% solids. Its unit O&M cost is $36 per wet ton or $211 per dry ton. Los Angeles' annual sludge quantity is 154,752 wet tons or 37,140 dry tons at 23 percent total solids. Its unit O&M cost is about $7 per wet ton or $27 per dry ton. (This facility has no cost for external bulking agents.)

The numbers above do not reflect revenues generated from the sale of finished compost, which is about $2.50, $0.60, and $1.40 per wet ton at the Hampton Roads, Columbus, and Site II static pile facilities, respectively. Revenue at the Los Angeles windrow facility is equivalent to about $1.00 per wet ton. Hampton Roads sells about 5,600 cu. yd. of compost per year, or about 60 to 70 percent of its annual production. Site II sells about 33,000 cu. yd. per year or about 34 percent of its annual production.

The EPA Report, Composting Municipal Sludge: A Technology Evaluation is available from the National Technical Information Service, Springfield, VA 22161. 175 pp. $18.95 paper, $6.50 microfiche. Listed as EPA 600/2-87-021 NTIS #PB87 17/252/AS.

Composting Economics

Another EPA report, *Estimating Sludge Management Costs,* was issued by the Agency's Office of Research and Development in October 1985. It was prepared by SCS Engineers in Long Beach, California, and was designed to assist in obtaining preliminary cost estimates for 34 common municipal wastewater sludge management processes. Preliminary base capital costs and base annual operation and maintenance (O&M) costs are calculated through the use of curves developed for each of the 34 processes. These curves are based on cost algorithms, which are outlined in the manual's appendices. The cost curves allow the user to rapidly obtain approximate cost estimates for sludge management processes based on only one or two process variables, e.g. annual sludge volume and distance hauled from treatment plant.

In preparing the cost curves, average default values were assumed for most of the variables contained in the cost algorithms. The majority of the algorithms are quite complex, says the manual's authors, having more process variables than the curves, which gives the user greater flexibility to adjust to site-specific characteristics. Therefore, the manual recommends using the curves to quickly obtain approximate costs for preliminary evaluation and to use the cost algorithms when more accurate site-specific cost estimates are needed.

The cost curves and algorithms for each process generally cover a range up to 100 million gallons of sludge per year. This range is approximately equivalent to a wastewater treatment plant of at least 50 mgd. The range was selected to include plants where supplemental cost information might be most useful.

Annual O&M component curves, presented for most processes, include annual requirements for labor hours, electrical energy, fuel, chemicals, maintenance materials and other items as needed. Base capital cost curves include mechanical equipment, concrete, steel, electrical and instrumentation, and installation labor.

Design parameters presented in the manual are "typical values" intended to guide the users in their efforts to obtain preliminary cost estimates. The more accurate the design information used, the more accurate the data that will result. Before using the cost curves or the algorithms, emphasize the authors, it is essential to understand the user's guide section to avoid inaccuracies.

Cost algorithms and curves were developed by SCS Engineers for such processes as lime stabilization, composting, transport, land application/disposal and storage. The remaining sludge management processes were already contained in the Computer Assisted Procedure for the Design and Evaluation of Wastewater Treatment Systems (CAPDET), the computer program used to do the estimates, although the SCS Engineers had to do some modifications.

SECTION THREE

END PRODUCT STABILITY AND MATURITY

11

USE OF MICROBIAL RESPIRATION RATE

George B. Willson and David Dalmat

THE ultimate test of the quality of compost is its effect on plants. However, field tests introduce a host of other variables and are time consuming and laborious. In order to make bioassays more useful, several scientists have used indicator plants. Spohn (1969) described a procedure to evaluate germination and growth of cress seed planted in beds of compost requiring 5 days. Zucconi et al. (1981) developed a test based on cress seed germination and root elongation in an aqueous extract from the compost. The procedure only requires 24 hours for incubation at 27°C and is quantitative. This method should have potential for monitoring quality of most composts. A disadvantage of bioassays is that they are not useful for evaluating differences in stability in early stages of the process.

A number of tests to predict the quality of compost are based on chemical analysis. The C/N ratio is widely cited as an index of compost quality. High ratios will interfere with the availability of nitrogen to plants. The composting process biooxidizes the carbon, reducing the ratio. The test is useful for high initial C/N ratio composts, but has little significance for low initial C/N ratio materials such as sludges and manures.

Another test is based on the disappearance of starch, Lassin (1970). Starch is commonly present in compost feed stocks and is readily digestible by the microorganisms, but the test is not quantitative. DeVleeschauwer et al. (1981) and Hirai (1983) analyzed composts for complex organic compounds, mainly organic acids. Several of these compounds are toxic to plants and are present in compost feed stocks or are intermediate decomposition products. DeVleeschauwer suggested using acetic acid as an indicator compound and correlated acetic acid concentration with germination test data. Usua (1983) described a test involving a 5-day biological oxygen demand of an aqueous suspension of the compost. These methods are potentially effective, but did require sophisticated analytical techniques. Other analyses that have been tested for possible application include: (a) nitrogen transformations, (b) redox potential, (c) sulfide test, and (d) chemical oxygen demand.

Desirable criteria for the test include: (a) the test should be quantitative over a wide range of conditions, (b) test sample size must be adequate to be representative of composts which are frequently heterogeneous materials, (c) the procedures should be simple to perform, equipment should be reliable and produce results quickly, preferably less than one day, (d) equipment and supplies for the test procedure should be inexpensive and readily available, (e) test data should correlate with bioassays, and (f) the procedure should be compatible with a variety of composting materials. Measurement of respiration rate was selected as having the most potential for meeting the criteria necessary for a widely acceptable stability test.

The procedure that seemed to meet most of the above criteria involved determining oxygen consumption rate by measuring the decrease in partial pressure of the oxygen in a sealed vessel containing compost. It is functionally similar to the Warburg test with the added stability of the Barcroft differential respirometer.

Measured respiration rates show a good correlation with processing time and procedure, thus the test is a measure of stability as it has been defined. The test meets the criteria suggested as important for acceptance as a measure of stability. It has the precision to detect small changes in processing time or procedure so should be suitable for process monitoring and quality control applications. Further correlation of respiration rates with bioassays is needed to establish the critical stability levels for good plant growth under field applications and horticultural uses and to investigate possible differences in critical values for different composts. Refinements of the test procedure have been suggested to improve its utility as a research tool.

A 75 percent decrease in respiration rate was observed during 30 days of aerobic curing. Aerobic curing was observed to be 6 to 8 times as fast as anaerobic curing.

The stability of compost is an important characteristic for predicting the potential plant response to use of the compost. There are, however, other compost characteristics that affect compost acceptability for use or influence plant response. Thus the respiration rate should be used in conjunction with measurement of other essential compost quality characteristics including: a) nutrient content; b) heavy metal content, c) pathogen content; d) carbon-nitrogen ratio, e) salts content; and f) pH.

References

DeVleeschauwer, D., O. Verdonck, and P. V. Assche. 1981. Phytotoxicity of refuse compost. BioCycle, Vol. 22, No. 1.

Haug, R. T. 1980. Compost engineering, principles and practices. Ann Arbor Science Publishers Inc., Ann Arbor, Michigan.

Hiria, M. F., V. Changyasak, and H. Kubota. 1983. A standard measurement for compost maturity. BioCycle, Vol. 24, No. 6.

Lossin, R. D. 1970. Compost studies, part I. Compost Science. Vol. 11, No. 6.

Niese, G. 1963. Experiments to determine the degree of decomposition of refuse compost by its self-heating capability. International Research Group on Refuse Disposal. Information Bulletin No. 17.

Parr, J. F., G. B. Willson, R. L. Chaney, L. J. Sikora, and C. F. Tester. 1978. Effect of certain chemical and physical factors on the composting process and product quality. *In* Proceedings of National Conference on Design of Municipal Sludge Compost Facilities. Hazardous Materials Control Research Institute. Silver Spring, Maryland.

Pressel, F. and W. Bidlingmaier. 1981. Analyzing decay rate of compost. BioCycle, Vol. 22, No. 5.

Spohn, E. 1969. How ripe is compost? Compost Science, Vol. 10, No. 3.

Umbreit, W. W., R. H. Burris, J. F. Stauffer. 1964. Manometric techniques, Fourth Edition, Burgess Publishing Co.

Usua, T., A. Shoji, and M. Yusa. 1983. Ripeness index of wastewater sludge compost. BioCycle, Vol. 24, No. 1.

Walkenshaw, G. B., D. Goodridge, D. Bowersox, and R. F. Noland. 1983. Columbus compost quality test. *In* Proceedings of the 15th National Conference on Municipal and Industrial Sludge Utilization and Disposal. Hazardous Materials Control Research Institute, Silver Spring, Maryland.

Zucconi, F., M. Forte, A. Monaco, and M. deBertoldi. 1981. Biological evaluation of compost maturity. BioCycle, Vol. 22, No. 4.

12

TEMPERATURE, PATHOGEN CONTROL AND PRODUCT QUALITY

Marco de Bertoldi, Franco Zucconi and M. Civilini

THE utility of composting is lessened by two potentially important risks or dangers associated with the end use of the product. The drawbacks are: 1) possibility of releasing heavy metals into the environment; and 2) the danger to human and animal health posed by human and animal pathogens that may also be present in the product.

Sludge and MSW may contain huge populations of pathogenic microorganisms—bacteria, viruses, fungi, and parasites. Some pathogens, such as parasites and viruses, cannot reproduce outside of a host, and as such may survive for a long time in a waste without increasing in number. By using assimilable matter in the waste as substrate, other pathogens (e.g., bacteria and fungi) may reproduce and gain in number with the passage of time. This latter group makes it necessary to provide conditions during the composting process such that the microorganisms not only are reduced in number, but also are unable to grow and multiply in the mature compost.

When the compost product is applied to the soil, its pathogenic populations are exposed to the homeostatic properties of the soil and to competition with microorganisms indigenous to the soil. The result is a further reduction in the number of pathogens. Despite the reduction, there is the risk that some of the pathogens may reach man in sufficiently high numbers as to cause disease or food poisoning. Hence, it is essential that the number of surviving pathogens in the compost product be sufficiently low as to impose a minimum or even no risk to health and hygiene.

Technically speaking, the necessary reduction could be accomplished through sterilization of the product. Unfortunately, sterilization not only is economically impractical, it also would not completely eliminate the risk of the regrowth of some pathogens (e.g., salmonellae). In the absence of competition, the latter could grow beyond measure. Furthermore, complete sterilization would not be necessary because the soil, which almost invariably is the destination of the compost product, already harbors certain pathogens (e.g., *Clostridium tetani, Aspergillus fumigatus*). It would be a superfluous undertaking and a waste of money to attempt to eliminate pathogens naturally present in soil.

Because of this situation, composting must guarantee that all pathogenic microorganisms not indigenous in the soil and posing a danger of contamination at sufficiently high concentrations be reduced to a level that eliminates danger. Otherwise, there is a risk of direct transmission of disease to humans who consume the contaminated food crops or of indirect transmission through the consumption of animals fed on contaminated feedstuffs.

There are two main problems associated with the reduction of pathogens: 1) the extent to which the sanitization of the composting material should proceed; and 2) the method or methods of determining the degree of sanitization completed. The answer to the first problem is a matter of economics. Waste, by definition, must not be burdened with an "overhead" in the form of cost of transformation into compost that would make the end product non-competitive on the open market. Consequently, extent of pathogen reduction must be a compromise between cost of processing and benefit accruing from a safe, sanitized end product. Degree of sanitization can be determined either during or at the completion of the composting process.

Sanitization is accomplished in composting by the high temperatures characteristically attained and maintained in a composting mass. Level of temperature and length of its duration must be selected on the basis of a logarithmic reduction of the principal pathogens. The degree of the reduction that should be attained is as yet to be defined. Nations have individually established temperature standards ranging from 55°C to 65°C covering a time-span ranging from 24 hours to 3 days.

Before being marketed, the compost product should undergo a series of analyses to determine the degree of its sanitization. Some laboratories base this determination on the disappearance of particular groups of pathogens, acting on the assumption that other pathogens simultaneously disappear. Others rely upon the use of indicator microorganisms. Experience thus far shows that total coliforms, fecal streptococci, enterobacteriaceae, certain viruses, and parasite ova can serve as satisfactorily reliable indicators.

For an indicator to furnish reliable information, it must satisfy four requisites:

1) It must always be present in sufficiently great numbers in the raw material to be composted.
2) It must belong to a group that has the same reactions to the treatment (i.e. thermal changes).
3) As a safety margin, it must have a resistance greater than that of the pathogens to be eliminated.
4) Tests involved in determining its identity and number must be simple and inexpensive.

One of the most controversial problems under discussion is the degree to which indicators must be reduced in order to have a safe level of sanitization in the compost.

The aim of this present research was to study several different composting methods, analyzing raw materials and finished products for the presence of a particularly important pathogen—*Salmonella sp.* and two indicators—total Coliforms and fecal Streptococci—which seem the most suitable of all for this job.

Growth Factors And Pathogen Requirements

Only certain pathogenic fungi and bacteria can multiply in compost, and only under certain circumstances. Although viruses and parasites can survive for a long time in the raw material to be composted, they cannot multiply outside of their respective hosts. The subsequent paragraphs and sections discuss the principal factors that affect the growth and survival of pathogens in compost.

Organic Matter: Because all pathogens in a waste are heterotrophic, their survival is dependent upon the availability of organic matter that can be metabolized by them. The molecular composition of the organic matter in fresh waste—and more so in sludge—is extremely heterogeneous. Included are substances that range from short-chain organic acids to highly complex molecules such as lignin. This diversity results in a good substrate for a wide variety of microorganisms. Thus, bacteria and pathogenic fungi generally can metabolize readily assimilable organic matter such as the simpler alcohols, or organic acids, sugars, etc., whereas they cannot multiply on complex compounds such as cellulose, lignin, and humic compounds. This limitation places the pathogens in an unfavorable competitive position with respect to the non-pathogenic microorganisms indigenous to the waste. In the composting process, a gamut of non-pathogenic organisms transform the organic matter of raw waste into simpler compounds through mineralization and humification. In cured compost, the residual fraction consists mainly of stable polymers, often combined to make up a complex humic fraction. Through the metabolic activities of the indigenous bacteria and fungi, readily soluble substances undergo the transformation that constitutes composting. The upshot is that the metabolism and growth of the pathogens is impeded by a shortage of assimilable organic matter.

Moisture: The moisture level in waste usually is sufficiently high to support pathogen growth. A very high moisture content, such as that in sludge, does not impair pathogen development because many pathogenic microorganisms, particularly the bacteria, are facultative anaerobes. At moisture levels lower than 25 percent, all microbial growth, including that of pathogens, slows and eventually ceases. Consequently, conditions prevailing in cured, stabilized compost would not be conducive to a regrowth of pathogens—provided that the moisture content remains lower than 30 percent.

Temperature: All pathogenic microorganisms have a threshold resistance that varies from one group to another and with environmental conditions. Sporogenous and nonsporogenous can survive exposure to temperatures above 100°C. Consequently, no compost process (system) can completely eliminate such microbial forms. Fortunately, the most dangerous of the pathogens in wastes, as far as the environment and agriculture are concerned, are neither sporogenous nor thermophilic. As such, they can be eliminated through a rather mild heat treatment.

Obviously, the effectiveness of heat treatment is a function of both the temperature attained and the length of exposure to that temperature. Moisture has the technical effect of increasing the conductivity of the exposed mass. Sanitization is brought about by high temperatures (e.g., 65°C) generated in the composting mass. The sanitization effect is greater if the moisture content is relatively high; and in fact, the lethal effect is much less pronounced if the material is relatively dry.

It is essential that the lethal temperature be reached simultaneously in all parts of the composting mass. Both requirements (all parts of the pile and simultaneity) are technically very difficult to achieve.

Oxygen Supply: Temperature rise in the composting mass represents the accumulation of heat released through exothermic reactions. Because the reactions are biooxidations, an adequate supply of oxygen is a requisite. Consequently, the attainment of high temperatures is possible only when sufficient oxygen is available for oxidation. In its absence, anaerobic processes take over and heat generation is sharply curtailed. The result is the failure of temperature to rise.

Microbial Competition and Antagonism: Microbial competition and antagonism rank among the most important of the factors in the control of pathogens in composting. The number of indigenous (i.e., "native" or "natural") saprophytes involved in composting is enormous. By contrast, the population of pathogenic microbes is numerically insignificant. This combination of a highly dense heterotrophic population and very small pathogenic population is characterized by a relatively high degree of antagonism and intense competition for nutrients in which the minute pathogenic population is definitely the loser. The disadvantage is intensified by the fact that composting material is not the natural environment for pathogens.

Practical Compost Systems

A principal objective of this paper is the evaluation of several types of composting systems in terms of their effectiveness in pathogen destruction. The evaluation includes an analysis of raw materials (waste) and finished product for the incidence of the particularly important pathogen, *Salmonella sp.* and two indicator microorganisms—total coliforms and fecal Streptococci. These two indicators seem to be the best suited to the task.

Open Systems

Turned Pile:

Turning alone does not ensure the attainment of a consistent or satisfactory oxygenation. Within an hour after the application of turning, oxygen levels within a pile drop drastically, and microbial bio-oxidative activity is correspondingly reduced. Consequently, to ensure adequate oxygenation, the pile must be turned frequently. However, providing the requisite frequency of turning leads to problems of a technical and economic nature.

Among the critical points to keep in mind about the turning operation is the importance of periodically exposing material from the outer layer of the pile to the conditions prevailing in the interior of the pile. Pile size is another important consideration. Thus, piles higher than about 3 m. are difficult to aerate to the extent needed for satisfactory sanitization.

Five experiments were conducted involving the turned pile approach. The raw material was the organic fraction of MSW obtained from the Dano plant at Pistoia (Italy). This organic MSW was mixed with aerobically stabilized sewage sludge. The piles were 1.5 m. in height, sheltered by roofing, and turned twice each week.

In two of the five experiments, Salmonella persisted longer than 30 days. A consistent drop in number of fecal coliforms and fecal Streptococci was not observed in any of the experiments. Repeated turning of the mass disturbed the composting process and altered the temperature profile. Nevertheless, oxygen needs in the center of the pile were such as to require turning at least twice per week.

Static Pile:

Bottom Suction: This is a static pile in which air is drawn through the pile by the imposition of negative pressure (bottom suction). With such piles, height is a critical factor. When they are greater than 2.5 to 3 m., uniform aeration becomes almost impossible. In a pile that is excessively high, the flow of air leads to the development of passageways ("short-circuiting") and sizable sections are not aerated. Because uniformity and freedom of airflow are functions of the porosity of a pile, the addition of materials such as wood chips helps to raise the permissible pile height.

The entire pile must be blanketed with an insulating layer (usually of cured, sanitized compost). This ensures that temperatures lethal to pathogens will be reached in all layers of pile—from the innermost to the outermost.

Research has shown that such systems perform fairly satisfactorily if properly operated (i.e., airflow not so great as to cool the pile or to dry the material too soon). In our research, we worked with piles, each of which had 20 tonnes of material, and was 1.5 m. high and 2.5 m. wide. In two trials, the raw material consisted entirely of the organic fraction of MSW; in another two trials, the raw material was a mixture of the organic fraction of MSW and aerobically stabilized wastewater sludge.

"Bottom Blowing": Forced aeration by blowing air through the pile (exertion of positive pressure) is a second form of aeration for static piles. Such a method of aeration tends to cool and dry the bottom layers of the pile, leaving the outer layers warm and moist. However, with careful operation, a well-sanitized compost can be produced.

Alternate Ventilation: As the name suggests, with such a system, bottom blowing is alternated with bottom suction. The periodic reversal of direction of airflow leads to an evenness of temperature level and moisture content throughout the pile. The result is a greater uniformity of sanitization.

In our experiments, an arrangement was used that involved two temperature probes and a solenoid valve adjusted to automatically reverse the direction of the airflow as soon as the difference between the temperature at a probe positioned 20 cm above the bottom of the pile and that at a second probe positioned 1.20 meters above the bottom exceeded 15°C. Findings showed that of the three approaches, best results were obtained with the alternate ventilation system.

Closed (In-Vessel) Systems

Continuous Vertical Reactors:

Several types of vertical reactors have a common characteristic of processing large quantities of material (as much as 2,000 m^3) in reactors that may be as high as 9 m. Material usually is loaded into the reactor through its top and eventually is discharged from its bottom. Oxygenation is provided by forcing air up from the bottom or through perforations in the bottom floor and through the composting mass. Conditions in these reactors are such that bottom ventilation is not sufficient to achieve the desired effects.

The attainment of optimum oxygenation levels requires a particular rate of air flow. As a result, the mass must be no deeper (thicker) than 2 to 3 m. Because the flow of air in a vertical reactor is from bottom to top, the terms "depth of material," "thickness of material," and "height of material" become synonymous when applied to such reactors. Deeper (thicker) masses require a proportionally greater quantity of air per unit of surface area according to the height of the contained mass. Supplying this air leads to the development of hyperventilation in the lower portion of the reactor, where the need for oxygen is least. The hyperventilation has many drawbacks, not the least of which are excessive cooling and drying. On the other hand, in the upper portion of the reactor, where the raw material is introduced and consequently the need for oxygen is greatest, oxygenation is inadequate. The problem is aggravated by the fact that as the forced air rises through the mass, its composition changes because of microbial action taking place in the reactor. Thus, CO_2 concentration gradually increases and O_2 concentration decreases. Consequently, the concentration of O_2 in the air, by the time it reaches the upper levels of the composting mass, is too low to support aerobic microbial metabolism. The usual consequences then occur: development of anaerobiosis and failure of the temperature to reach levels lethal to microbes, and the production of an inadequately sanitized end product. The occurrence of partially anaerobic and mesophyllic conditions is ideal for the proliferation of Enterobacteria and Streptococci.

Inadequate ventilation—regardless of direction of airflow (top to bottom, or bottom to top)—and consequent failure to reach lethal temperatures in the upper levels are the most serious disadvantages in the use of the continuous vertical reactor in which the composting mass is 7 to 10 m. high. This assessment was confirmed by our analyses of material obtained from a 100-m^3 vertical reactor charged with a mixture of sludge and organic MSW.

Horizontal Reactors:

As would be surmised from the name of the reactor, the composting material is arranged along the length of the unit. The depth of the mass does not exceed 2 to 3 m. The orientation and depth of the composting mass is comparable to that with the static pile method. The advantage of the horizontal reactor is the ability to control the course of the compost process. Because oxygen is supplied either by turning or by aeration—or both—the composting mass can be uniformly oxygenated and the temperature can be readily controlled, and a satisfactorily sanitized product can be produced in quantity.

Conclusions

Pathogenic Bacteria and Indicators:

A major challenge with the marketing and use of compost is the possibility of pathogenic microorganisms being present in the product. Indiscriminate use of non-sanitized compost products could result in the dispersion of pathogens in the environment, and thereby give rise to a serious threat to the health and well-being of man and animal. Therefore, all such products must be treated before being released for utilization. The treatment should be such that the incidence of all pathogens in the products be reduced to a level lower than the risk level. The attainment of such a goal on a practical scale must be preceded by the resolution of two problems: 1) Determination of the threshold level below which pathogens no longer constitute a hazard; and 2) development of a program of monitoring the compost product with respect to sanitization.

Finding an answer to the determination of a threshold level is not an easy task since the level varies from one organism to another. Moreover, the determination is constrained by the practical fact that "there is no such thing as an economically feasible, completely hazardless level." With respect to monitoring, it must be admitted that the decision as to which pathogen to monitor is open to question at present. This research attempts to make a contribution in that direction.

In view of the wide array of pathogens that could be present in the raw waste, a thorough analysis of the entire compost output for its pathogen content would become an enormous economic problem. Therefore, it would be more practical to select a few representative pathogens that are easy to identify and quantify and are likely to be present in most raw materials (wastes). The genus, *Salmonella,* would meet these specifications. The presence of Salmonella in a product would indicate poor sanitization.

To avoid the necessity of analyzing for all types of pathogens, recourse could be had to the use of indicator microorganisms. Our results, as well as those obtained by others in research on the use of fecal coliforms and fecal streptococci as indicators, especially of the presence of Salmonella, indicate that the two indicators are abundant in raw wastes (although Salmonella may not be found at times). The fecal coli and fecal streptococci survive without difficulty under aerobic and anaerobic conditions. However, coliforms are more sensitive to temperature (as is Salmonella), whereas streptococci are more resistant (45-48°C). In all composting trials, we found that Salmonella disappeared before the coliform level dropped to 5×10^2, and streptococci dropped to 3×10^3. Both indicators can multiply in waste under favorable conditions. Hence, both can be used for indicating regrowth.

At this stage of analysis, it would be interesting to extend the research to include a determination of the limits to be established for the two indicators, and to investigate their potential as indicators of the presence of pathogenic viruses and parasites. Our studies found that if the temperature of the composting mass reaches 65°C and remains at that level for two days, *Salmonella sp.* disappears completely. Moreover, under those conditions, the concentration of coliform indicators is always less than 5×10^2; and that of Streptococcus, less than 3×10^3. Therefore, those two concentrations could be used as indicators of the attainment of satisfactory sanitization—at least with respect to pathogenic bacteria.

Compost Processing and Sanitization:

Composting and sanitization are not necessarily synonymous. For sanitization to take place, certain conditions must be supplied during the composting process. In theory, the end product can justifiably be considered sanitized if the compost process has been properly conducted and the entire mass has been exposed to lethal temperatures. In practice, meeting those two requirements is not enough to guarantee the necessary "stability" with regard to sanitization, inasmuch as pathogens may "regrow." Consequently, composting must not only drastically reduce the number of pathogens, it must also transform the waste such that regrowth can not occur. This can be done by adequately stabilizing the organic matter, mineralizing all simple compounds readily assimilable by pathogens, and by humifying other compounds. Moreover, the moisture content of the product should be fairly low (25-30 percent).

Although many factors are involved in the elimination of pathogens, temperature is the only one that can be quantitatively measured and controlled. Thus, it should be used to control the process. Furthermore, since rise in temperature is the direct result of the bio-oxidation activity of the microbial populations, conditions promoting the bio-oxidation should be provided.

The conclusion to be drawn from the preceding two paragraphs is that from a practical standpoint, satisfactory sanitization is a function not only of the method or system used in composting the waste, but also of the conditions prevailing during the composting. Thus, merely turning the waste periodically without supplying ventilation does not guarantee good sanitization—unless the piles are low (1-1.3 m.). Conditions in the usual continuous vertical reactor (9 m. column of contents) are not conducive to the production of a sanitized product. On the other hand, successful results were the rule when ventilated piles and horizontal reactors were used in this research.

The most important parameters with respect to attaining sanitization are the following: 1) The entire mass must be maintained at a minimum of 65°C for two to three consecutive days; 2) the material must be biologically stable so as to prevent pathogen regrowth. Consequently, to be able to decide with a reasonable degree of certainty that a given compost product is hygienically safe, it is necessary to know the precise conditions that prevailed throughout the composting process whence it came.

Marco de Bertoldi is Professor in Industrial Microbiology, Faculty of Agriculture, University of Udine, Italy. Franco Zucconi is a professor in the Faculty of Agriculture, University of Naples. Prof. M. Civilini is in the Faculty of Agriculture, University of Udine. *This paper was supported by a contract with the Commission of European Communities, Directorate-General Science, Research and Development Programme on Recycling of Urban and Industrial Waste.*

13

MATURITY EVALUATION OF ORGANIC WASTE

A. Saviozzi, R. Levi-Minzi and R. Riffaldi

To be compatible with agricultural uses, a compost must have been sufficiently stabilized, i.e., it must have been decomposed to a humus-like product. The reason is that if incorporated in the soil, an immature compost can cause severe damage to plant growth. Because stabilization occurs within a varying period of time, one or more practical indexes for determining the state of composting progress and the maturity attained are in great demand.

Several methods have been proposed for determining the degree of compost stability. Among the many proposed, Harada and Inoko[1] suggest the measurement of the CEC, and Harada *et al*[2] suggest total N content, the C/N ratio, and the ratio of carbon in reducing sugars to total carbon. Zucconi *et al*[3] favor the "germination index" in which *Lepidium sativum* L. is the test plant. Regarding the water extract from compost, Yoshida and Kubota,[4] Chanyasak *et al*,[5] Hirai *et al*,[6] respectively, proposed the elution patterns of gel chromatography on *Sephadex C-15* as a measure of stability. The same criteria had been used previously for the water-extract characterization of the stabilized product from the composting of paper-processing wastewater sludge with straw[7] and from the composting of farmyard manure (FYM). A review of various proposed indicators of maturity of municipal refuse compost was recently published by Morel *et al.*[9]

Among the chemical criteria required for the characterization of an organic waste stabilization product (e.g., compost), humus content and quality rank high. This aspect was investigated by several authors,[10-15] each following a different experimental approach.

The objectives of this study were to: 1) attempt to follow the course of the maturation process in composting through the use of various humification parameters; and 2) to evaluate suitability of the parameters for statistically assessing the maturity of different organic wastes.

Even though the C/N ratio may be relatively low at the initial stage of composting, it is commonly used as a criterion for determining the degree of maturity of many kinds of compost.[11,15] For this reason, simple linear correlation analyses were performed between the various humification parameters and the C/N ratio. All of the parameters are significantly related to such a ratio. Of them, the humification index, C_{HA} (percent of organic C), described by Roletto *et al*[12] shows the highest level of correlation (i.e., $P = 0.01$).

Because the parameters described above proved to be suitable, they were used in the assessment of the changes that occurred in the organic matter of FYM over a 3-month stabilization period. This was done by comparing the characteristics of samples collected on day-1 with those of samples collected after three months of stabilization (storage). The significant difference between fresh and mature FYM as defined by the paired t-test analysis (cf. Table 1) indicates the suitability of the various indexes for evaluating the maturity of FYM samples. A comparison between the values presented in Table 1 and those reported by Roletto *et al* for a mature FYM[12] shows that the

TABLE 1.

Statistical Analyses Of Some Humification Parameters Of The Six FYM Series.

	Org. C %DM	HI	C_{HA}/C_{FA}	C_{HA+FA} %Org.C	C_{HA} %Org.C	C/N
			Fresh samples			
Range	38.7-40.1	0.5-1.1	0.6-1.1	10.6-14.5	4.4-7.5	16.8-21.9
Mean	39.2	0.8	0.8	12.3	5.4	19.0
CL	0.6	0.3	0.3	1.7	1.1	1.9
CV	1.3	24.4	23.8	11.3	20.2	9.5
			Mature samples			
Range	34.8-37.8	0.2-0.4	1.1-3.0	16.5-21.4	9.7-13.8	10.6-16.9
Mean	36.0	0.3	1.9	18.2	11.6	14.2
CL°	1.1	0.1	0.8	2.2	1.9	3.1
CV^	3.0	24.5	35.5	10.3	5.2	19.2
t"	−6.2	−5.7	5.0	4.9	9.9	−2.8
r-C/N	0.75	0.74	−0.60	−0.65	−0.65	—

°CL = confidence limit at P=0.05 probability level.
^CV = coefficient of variation (%).
"t = differences between fresh and mature sample values evaluated by the paired t-test analysis.

TABLE 2.

Statistical Analyses Of Some Humification Parameters Of the Four Urban Refuse Composts.

	Org.C %DM	HI	C_{HA}/C_{FA}	C_{HA+FA} %Org.C	C_{HA} %Org.C	C/N
			Raw samples			
Range	27.0-33.1	2.9-3.5	0.2-0.5	7.6-10.3	1.4-3.0	21.5-29.4
Mean	30.2	3.2	0.3	8.9	1.9	25.9
CL	6.0	0.4	0.4	2.6	1.7	7.3
CV	9.5	7.9	47.1	14.0	37.8	13.0
			Compost samples			
Range	9.9-28.5	0.2-1.2	0.4-2.8	11.2-31.6	3.5-23.3	9.1-17.0
Mean	21.1	0.5	1.8	17.8	11.5	13.1
CL	17.2	0.9	2.1	20.2	18.5	7.7
CV	37.5	87.1	59.1	53.3	74.6	27.6
t	−3.3	−9.1	3.0n.s.	1.8n.s.	2.2n.s.	−14.1
r-C/N	0.81	0.86	−0.84	−0.75	−0.80	—

values obtained in this study regarding the humic fraction are, on the average, higher than those reported by Roletto *et al* for: 1) the polymerization ratio C_{HA}/C_{FA}, 2) the C_{HA+FA} (percent of organic C) defined by Roletto *et al* as the humification ratio, and 3) the humification index C_{HA} (percent of organic C). The differences in values probably are due to the different extractants and procedures. Furthermore, the mean value of the index of humification (HI) described by Sequi *et al*[13] is 0.3—a value that is in agreement with the one reported for a well matured cow manure by the authors of this paper.

Of the various parameters that have been used in the assessment of the degree of humification, the HI shows the highest correlation with the C/N ratio (i.e., P = 0.01). Nevertheless, all of the indexes are well related to such a ratio, even if it be at a lower significance.

As can be seen from the data in Table 1, an analysis of the data by the confidence limit test allows us to suggest minimum values for some humification parameters for maturity of FYM. An FYM containing less than 37.1 percent of organic carbon, and having an HI lower than 0.4, a C_{HA+FA} (percent of organic C) higher than 16, and a C_{HA} (percent of organic C) higher than 9.7 could be regarded as being mature.

In Table 2 are reported the statistical values for humification parameters related to urban refuse composts. A comparison between composted refuse and mature FYM samples shows that, although the mean value of the total organic carbon in composted municipal refuse is about 40 percent lower than that in the mature FYM, the averages for the various humification parameters are quite similar. However, due to the higher coefficient of variation and probably to the lower number of samples examined, the paired t-test analysis for compost shows that only the carbon content, C/N, and HI are significantly different between raw and compost samples.

The HI value 3.2 for raw refuse and 0.5 for composted refuse taken in conjunction with the HI 2.52 reported by Sequi *et al*[13] for a partially matured sample would seem to indicate that the HI is suitable for the characterization of the humified matter content in composted urban refuse. Furthermore, the HI expressed as C_{HA}/C_{FA} shows the best correlation with the C/N ratio (i.e., $P = 0.01$), although the other indexes also show a significant correlation with the ratio. Analysis of the data by the confidence limit test shows that the only minimum value that can be derived for evaluating the degree of maturity of compost is that of 1.4 for HI.

The following conclusions were developed: 1. It seems to be possible to evaluate chemical stabilization of organic wastes on the basis of various humification indexes determined in the laboratory, i.e., a place in which controls can be applied more rigorously than in the field. 2. Among the parameters tested, the HI seems to be the most appropriate, even though it is less simple and rapid. 3. Before defining which among the parameters is the most suitable for evaluating the degree of maturity of organic wastes and utility in crop production, a much greater number of wastes of the same kind but composted under different conditions must be tested.

References

1. Harada, Y. and Inoko, A.: "The Measurement of the Cation-Exchange Capacity of Composts for the Estimation of the Degree of Maturity". *Soil Sci. Plant Nutr., 26,* 126 (1980).

2. Harada, Y., Inoko, A., Tadaki, M. and Izawa, T.: "Maturing Process of City Refuse Compost During Piling". *Soil Sci. Plant Nutr., 27,* 357 (1981).

3. Zucconi, F., Forte, M., Monaco, A. and de Bertoldi, M.: "Biological Evaluation of Compost Maturity". *BioCycle.* July/August, 27 (1981).

4. Yoshida, T. and Kubota, H.: "Gel Chromatography of Water Extract from Compost". *J. Ferm. Technol., 6,* 582 (1979).

5. Chanyasak, V., Yoshida, T. and Kubota, H.: "Chemical Components in Gel Chromatographic Fractionation of Water Extract from Sewage Sludge Compost." *J. Ferm. Technol., 58,* 533 (1980).

6. Hirai, M.F., Chanyasak, V. and Kubota, H.: "A Standard Measurement for Compost Maturity". *BioCycle,* November/December, 54 (1983).

7. Saviozzi, A., Riffaldi, R. and Levi-Minzi, R.: "Compost Maturity by Water Extract Analysis". In: Compost: Production, Quality and Use. (de Bertoldi, M. et al. Eds.), Elsevier Applied Science, London, 359 (1987).

8. Riffaldi, R., Saviozzi, A. and Levi-Minzi, R.: "Water Extracts of Fresh and Mature Farmyard Manure". *Biol. Wastes, 23* (1988) (in Press).

9. Morel, J.L., Colin, F., Germon, J.C., Godin, P. and Juste, C.: "Methods for the Evaluation of the Maturity of Municipal Refuse Compost". In: Composting of Agricultural and Other Wastes (Gasser, J.K.R. Ed.). Elsevier Applied Science, London, 56 (1984).

10. Kuwatsuka, S., Tsutsuki, K. and Kumada, K.: "Chemical Studies on Soil Humic Acids. I. Elementary Composition of Humic Acids". *Soil Sci. Plant Nutr., 24,* 337 (1978).

11. Sugahara, K. and Inoko, A.: "Composition Analysis of Humus and Characterization of Humic Acid Obtained from City Refuse Compost". *Soil Sci. Plant Nutr., 27,* 213 (1981).

12. Roletto, E., Barberis, R., Consiglio, M. and Jodice, R.: "Chemical Parameters for Evaluating Compost Maturity". *BioCycle,* March, 46 (1985).

13. Sequi, P., De Nobili, M., Leita, L. and Cercignani, G.: "A New Index of Humification". *Agrochimica, 30,* 175 (1986).

14. Levi-Minzi, R., Riffaldi, R. and Saviozzi, A.: "Organic Matter and Nutrients in Fresh and Mature Farmyard Manure". *Agric. Wastes, 16,* 225 (1986).

15. Riffaldi, R., Levi-Minzi, R., Pera, A. and de Bertoldi, M.: "Evaluation of Compost Maturity by Means of Chemical and Microbial Analyses". *Waste Manag. and Res., 4,* 387 (1986).

16. Levi-Minzi, R. and Riffaldi, R.: "Chemical Differences Between Fresh and Composted Municipal Wastes". (submitted).

17. Schnitzer, M., Lowe, L.E., Dormaar, J.F. and Martel, V.: "A Procedure for the Characterization of Soil Organic Matter". *Can. J. Soil Sci. 61,* 517 (1981).

18. Lowe, L.E.: "Fractionation of Acid Soluble Components of Soil Organic Matter Using Polyvinylpyrrolidone". *Can. J. Soil Sci., 35,* 119 (1975).

19. Bremner, J.M.: "Total Nitrogen. Inorganic Forms of Nitrogen. Organic Forms of Nitrogen". In: Black, C.A. et al. (Eds.), Methods of Soil Analysis. ASA, Madison (1965).

The authors are on the staff of the Institute of Agricultural Chemistry, University of Pisa, Italy. *The research reported in this paper was supported by MPI, Italy. Grant 60%.*

14

SPECIFICATIONS FOR SOLID WASTE COMPOST

Franco Zucconi and Marco de Bertoldi

PROMOTING the recovery of organic wastes through composting requires a standard definition of process and product as a prerequisite for ensuring quality and fostering agricultural use. Indeed, the task of defining quality standards is far from simple, considering the existing heterogeneity of processes and plants, and the great diversity of materials that occur in city waste. Furthermore, specifications for organic soil conditioners are limited, and definitions are conflicting or lack specificity. Yet, a standard nomenclature is necessary to discriminate compost from products of different origins. This requires decisions on which definitions may reduce approximation, and which guidelines may help to improve process performance and product quality.

Accordingly, this chapter introduces a number of specifications on composting, stabilization and hygienic standards, generally related to minimum values of desirable constituents and to maximum tolerable loads for unwanted ones. It also discusses the state-of-the-art as a necessary means to account for the specifications introduced and for the existing limits in the identification of standards. The advantage of establishing specifications becomes apparent when one considers: a) The need for promoting industrial management of organic wastes and the reclaiming of energy; b) The urgency for introducing some order into the existing array of processes and marketed products so that composts may be identified and compared; and c) The requirement for products that are compatible with different agricultural uses.

The discussion that follows gives guidelines for the production and standard evaluation of compost originating from municipal solid waste (MSW), as well as from any mixture which contains MSW together with industrial or agricultural residues, and sludges from wastewater treatment in proportion appropriate to the solid waste. These guidelines apply strictly to the industrial production of marketed compost to be used on agricultural crops. The classification of composts refers to products derived at different phases of the stabilization process. These products range from "fresh" to "cured" compost.

The Definition of Composting: The term compost, as most often used, refers to the uncertainly-defined product of waste organic matter decomposition. Such a usage is too vague, however, for separating processes or for recognizing dynamic sequences that lead to a specific product. Therefore, it is necessary to restrict the term "composting" to a specified type of process according to the definition proposed: *Composting is a controlled biooxidative process that:* 1) *Involves a heterogeneous organic substrate in the solid state;* 2) *Evolves by passing through a thermophilic stage and a temporary release of phytotoxin; and* 3) *Leads to production of carbon dioxide, water, minerals and stabilized organic matter (compost).*

Key words and phrases:

a) *Controlled . . . Process:* Distinguishes "composting" from processes that occur in the absence of an imposed control of temperature, moisture, substrate composition and oxygenation.

b) *Biooxidative:* Distinguishes composting from physical and chemical treatments, as well as from other non-aerobic operational conditions.

c) *Organic Substrate:* Is a prerequisite for "biological stabilization" and the production of organic soil conditioner.

d) *Heterogeneous:* describes substrates containing a mixture of organic materials from different origins. MSW fits well in this category.

e) *Solid State:* Is consistent with the practice of organic waste treatment, well established in industrial processes.

f) *Thermophilic Stage:* Substantial quantities of heat are produced in the initial part of the process which account for the temperature rise. The rate of heat production declines rapidly during the subsequent stabilization stage (associated with a temperature drop below 40-35°C). This dynamic quality of heat production may be considered a unique condition for characterizing composting.

g) *Temporary Phytotoxin Release:* The metabolic production of phytotoxins characterizes the initial stage of fresh organic matter decomposition. Toxin production apparently is of shorter duration with heterogeneous substrates and under aerobic conditions. A persistent phytotoxicity indicates poor process development (often insufficient oxygenation) or poorly balanced substrates.

h) *Carbon dioxide, Water and Minerals:* Refer to some of the products of biooxidative degradation processes.

i) *Stabilized Organic Matter (Compost):* Composting by definition leads to a stabilized (humified) product that is highly valuable for agricultural uses (organic soil conditioner).

The Definition of Compost: Compost is the product of the composting process, having passed through a number of sequential changes. Accordingly, the composting process must also be defined in terms of stages leading to a specific product. To be properly called compost the organic matter must have been stabilized to a humus-like product, and degraded to fine particles, having lost its original identity. As a stable product, compost can be stored without further treatment, and can be applied on the land without damage to standing crops.

The term "stabilization," although commonly attributed to the whole process, would be more appropriate for the second stage of composting characterized by humification. This stage follows an initial "decomposition" stage where destabilization prevails. Heat and toxin production differ greatly between the two stages. They are at a maximum during decomposition, tending to subside as the process achieves stability. Consistently, a definition of compost may be given that links the product to a specific process and to a precise stage of its evolution: *Compost is the stabilized and sanitized product of composting which is beneficial to plant growth. It has undergone an initial rapid stage of decomposition and is in the process of humification (stabilization stage).*

Key words and phrases in this definition are:

a) *Stabilized:* Stabilization is a prerequisite for storage, marketing and use in agriculture. Stabilization must be achieved through the biological process itself and must not be confused with or substituted by desiccation. Although the latter may lead to an appearance of stabilization, the decomposition (latent metabolism) would resume when conditions again become favorable.

b) *Sanitized:* To be of practical application, the composting must result in a sanitation of the organic matter, and thus allow a direct utilization of compost in agriculture. Pathogen destruction is accomplished through the combined effects of high temperature and competition with a non pathogenic microbial population, also favored by the thermophilic condition. Sanitation through composting must not be confused with, or substituted by, artificial pasteurization or sterilization.

c) *Product of Composting.* It associates compost with the process, and excludes products from other processes.

d) *Beneficial to Plants:* It is essential that the product does not adversely affect plant viability.

e) *Initial Rapid Stage of Decomposition:* Defines "decomposition" as being the first of two stages of organic matter degradation.

f) *Process of Humification:* Defines the stabilization stage as being characterized by a process of humification. It also implies that the first stage must be terminated and the presence of soluble organic metabolites should be at a very low level.

The above definitions tie "compost" and "composting" to the controlled biooxidative treatment of organic wastes. The advantage of restricting the term "composting" to controlled biooxidative processes is apparent when the impact on product quality and the improvement of the treatment of waste organic matter are considered. Compared with other processes, controlled biooxidation ensures greater efficiency, is more effective hygienically, eliminates objectionable odors and shortens the time needed for stabilization. All of these conditions are essential to the promotion of sound industrial management of organic wastes.

Compost Specifications

Origin of Product. The term "compost" without qualifying adjectives must be restricted to the product of composting as defined above. Consequently, other names must be given to the products of fermentation or anaerobic digestion. To be marketed for agricultural use, compost must be accompanied by specifications relative to its origin, composition and degree of stabilization. When adding stabilized organic matter of different origin to compost for marketing, the mixture may not be designated as "compost": i.e. it should be described as *organic soil conditioner* and the description be qualified by reference to specific components.

Physical and Chemical Parameters. The characteristics and composition of compost vary according to the type of starting material. The data listed in Table 1 refer to values compatible with compost from municipal solid waste.

Other required characteristics are:

a) *Organic matter:* specifications should include % content of organic matter and the degree of its stabilization (see: compost classification). In relating organic matter to organic carbon, coefficients derived from soil analyses (K = 1.78) appear to be too small. A K value closer to 2.00 is suggested.

b) *Moisture content:* lowering of the moisture content is an indication of progress in the composting process. Low moisture content avoids unnecessary transportation of water, and helps in sieving, storing or spreading operations. Moisture content may vary according to the climatic conditions in which the process is done. Yet, the moisture content must be stated on marketed compost to help in determining its commercial value.

c) *Inerts:* substantially inert-free compost is required for most agricultural uses. Glass and plastics in particular need to be contained within the levels stated in Table 1. Ferrous metals are practically absent. Non-ferrous metals are generally present at a low level and may be separated following composting. Separation of inerts is relatively straightforward in well stabilized and dried products thus providing an indication on the completion of the composting process. This also differentiates composting from desiccation or similar processes where size reduction of organic matter did not occur thus making difficult its separation from inerts.

d) *Mineral content:* admissible levels for certain nutrients in compost are summarized in Table 2.

Nitrogen: total levels not less than 0.6% dry weight; organic nitrogen not less than 90%; inorganic nitrogen should mostly be in the nitrate form. The concentration of NH4+ in the finished compost should not exceed the level of 0.04% dw. (except for the "fresh compost"). The presence of NH4+ is an indication of unstabilized materials (decomposition stage, anaerobic processes.) It also tends to occur when starting materials have a low C/N ratio.

P.K.: normal range for P205: 0.5-0.9% dw.; K20: 0.2-0.8% dw.

Micronutrients: micronutrient requirements are difficult to assess. Certain micronutrients are essential for plant growth. The presence of Fe or Zn chelates is advantageous to plants.

e) *Initial and final C/N values must be declared:* C/N ratio must be below 22 in order to achieve the stabilization degree required for "compost" when the starting material has a C/N in the range of 35-40 or slightly above. When wastes contain nitrogen-rich products, the final C/N values are expected to be proportionally lower, yet not below 11-12, corresponding to humus values.

f) *Salinity:* an exceedingly high salinity may adversely affect plants and crops. Damage is proportionally greater in non-calcareous soils. Total salinity (conductivity should not exceed 2 g salt/liter (expressed as NaCl) and the concentration of Na and Cl ions should be specified in marketed compost.

g) *pH:* most common pH values range 6.5 to 8.0. Compatibility with plant growth is within the range of 5.5 to 8.0. If pH values are higher or lower, the reason for their departure from standards should be specified.

Hazardous Materials: Wastes contaminated with hazardous materials should not be used for the preparation of organic soil amenders. Hazardous materials of importance are heavy metals and slowly degradable toxic organic compounds. At present, the content of heavy metals in compost is increasing. The use of sludges in composting can add to the problem. Therefore, starting materials should be selected that will not contaminate the compost produced and pollute the soil. Pretreatments of solid wastes should be geared in order to restrict composting to the organic fraction and limit the transfer of heavy metals into it. To achieve this, inerts should be removed prior to composting.

The concentrations of heavy metals in compost may be referred to those given in the EC Directive for sewage sludge disposal to agricultural land. However, sludge disposal is essentially a water-disposal operation. Considering the high amount of organic matter annually requested for soil conditioning (5-25 t dry matter/ha) maximum permissible levels of heavy metals in compost should then be proportionally lower. A multiplication factor of about 0.3-0.2 is suggested for adapting to the compost specifications the EC levels on heavy metals set for sludge. This suggestion is consistent with the heavy metal content

TABLE 1

Compost Characteristics According to the Grade of Marketable Product.

Sieve Size	Maximum Inerts (% dry wt)					Maximum Moisture %	Biodegrad. Org. Matter (Minimum) % dry wt	
		Glass		Plastic				
Grade	mm	A	B	A	B	A-B	A	B
Very fine	8	1	0.1	0.4	0.2	30	20	30
Fine	16	2	1	0.8	0.4	35	25	35
Medium	24	4	2	1.6	0.8	40	30	40
Coarse	40	6	3	3.5	1.6	50	35	45

Values listed under columns A represent today's situation; Those of columns B represent target values at which to aim for the development of the process and product.

TABLE 2

Total Amounts Of Nutrients In Compost (Dry Weight Basis).

Nutrients	Admissible Levels (minimum)
N	0.6%
P205	0.5%
K20	0.3%
Ca0	2.0%
CaC03	3.0%
Mg0	0.3%

which may be achieved by early separation of the organic fraction from wastes or, better, by the separate collection of organic wastes to be composted.

Table 3 gives data on heavy metal content in compost, relative to conventional and separate collection of wastes. Compared to values of heavy metals in compost in Table 3, values of minimum load/year would appear rather high. However, the EEC proposal for sludge takes into account the reduced impact of heavy metals in soil due to insolubilization effects and to the relatively smaller assimilation by the plants. Values included in Table 3 should not apply in pot cultivation of ornamental plants, for which higher values may be permitted.

As with the heavy metals, hazardous organic compounds (detergents, pesticides, phenols, hydrocarbons, products of pharmaceutical industries, etc.) must be dealt with according to existing legislation on hazardous waste disposal. They pose a health risk either directly or through their metabolic (breakdown) products and not be allowed to be present in the compost produced for agricultural use. Again, less vigorous conditions may be permitted for the pot cultivation of ornamental plants.

Stabilization. To be compatible with agricultural uses, composts must be stabilized to a degree at which they do not hamper plant viability. To assess the extent of stabilization appears to be a difficult task, however, considering the present incomplete understanding of the parameters involved and related standards. An effort should be made to acquire more experience in a number of fields related to stabilization:

Humification: The stabilization stage is one of progressive synthesis and polymerization in which "humic" components are prevalent. Disappearance of soluble substrate and catabolites and especially the appearance of new humus-like polymers may provide a valuable measure for stabilization. The value of the analyses of humic components should be clarified, particularly when applied to products of different origins and processes, or when stabilized organic matter from other sources (root, natural humus, etc.) is added to the marketed compost. Consistently, it is suggested that humified organic matter should be at least 10% of organic carbon. Specific research should clarify whether this value could be increased.

Phytotoxicity: Metabolic toxins are produced throughout decomposition, tending to subside during the stabilization stage. Toxicity measured with biological tests may indicate the degree of degradation and fit with the need for assessing physiological compatibility with plants and the best use of compost. The degree of maturation required may vary according to compost classification and its specific uses.

A simple method which may be carried on in the factory is related to instantaneous toxicity (i.e. the toxicity which may be measured at any moment). The value of this test is bound to continuous monitoring of the com-

TABLE 3

Limiting Values For Heavy Metals In Relation To Concentration In Compost.

	Maximum In Compost (mg/Kg dry wt) standard collection[1]			Maximum Load Year[3] Kg/ha		Maximum In Soil[3] mg/Kg dry wt	
	R	M	separate collec.[2]	R	M	R	M
Zn	1000	1500	240	25.0	30.0	150	300
Pb	750	1000	160	10.0	15.0	50	100
Cu	300	500	40	10.0	12.0	50	100
Cr	150	200	30	10.0	—	50	—
Ni	50	100	10	2.0	3.0	30	50
As	—	—	—	0.35	—	20	—
Hg	5	5	0.5	0.40	—	2	—
Cd	5	5	1	0.10	0.15	1	3

Annual addition and content in agricultural soils to which compost is applied. R and M values refer, respectively, to recommended and mandatory limits.

[1]Flemish Public Waste Company (VAM) limits for compost use in Agriculture: requested maximum levels relative to use on food crops (R in this case) or on ornamentals (M).

[2]Results of separate collection of degradable organic fraction of MSW in de Bilt, Netherlands (VAM).

[3]Proposal for the EEC Council Directive (25-5-1984) for the use of sludge in agriculture.

posting process and understanding of the various phases through which it evolves.

Latent metabolism: The stabilization of organic matter is accompanied by a number of metabolic changes. Incomplete stabilization, or interruption of composting, leads to a latter resumption of metabolic process. Dynamic analyses of respiration, heat and toxin show a strict relationship between the metabolic phase already achieved and the complementary (latent) evolution occurring when conditions again become favorable. Laboratory analyses on latent metabolism then have the potential for providing objective information on the stabilization of organic matter, independent of material origin or our knowledge of the processes undergone. This allows the determination of how far the process was carried out, even on unknown products.

Available methods for latent metabolism are based on the evaluation of oxygen uptake, self-heating capacity (i.e. production of heat in controlled conditions), or toxin evolution (latent toxicity).

Other parameters:

C/N ratio: C/N ratio undergoes changes during composting and yet its dynamics strongly vary with different substrates. A decline from an initial C/N of 35-40 or higher to a final level of 18-20 would imply an advanced degree of stabilization. However, care must be taken in interpreting the significance of the C/N level in composts of unknown origin. When products rich in nitrogen serve as starting substrates (e.g.: animal wastes, sludges from water treatment, etc.), the initial C/N ratio may be low (even below 10) and tends to rise during the process of stabilization. Therefore, it is necessary that both initial and final C/N values be indicated for a correct evaluation of stabilization. Also, a concerted effort should be made to acquire more experience in this field and to define changes which may help in assessing stabilization.

Chemical or biological oxygen demand: A number of suggestions have been made that BOD or COD values should be introduced. Before such values are made a part of specifications, however, the manner must be defined in which COD and BOD are related to the process evolution and stabilization.

Microbiological Specifications for Sanitized Compost: Composts have to meet the requirements of epidemiology hygiene. This can be achieved if pathogenic organisms have been rendered harmless for man, animals, and plants whether there is need to take cognizance of the end use of compost. When compost is used for ornamental plants, which are often kept in apartments, the absence of human pathogens should be ensured. Less severe requirements are needed for its use in open fields and forestry. Composts should be assessed in terms of hygiene by testing both the composting process and the end-product.

1. *Processes:* Many microbiological investigations of composting plants have shown that the requirements of hygiene are met under proper operation. The chemical and physical parameters of the different processes have to be considered, together with the differing types of processes. The sanitization effect depends on the duration and temperature of the decomposition process while temperature itself depends on chemical and physical parameters, such as water content, aeration, pH-value, particle size, available organic matter. For efficient disinfection of compost, the following requirements should be met:

a) *Windrow composting.* To ensure a hygienically safe product, the composting process should exceed 65°C for at least one week. It is most important that all of the compost mass in the windrow reaches this temperature. These requirements can be met by operating a program of regular turning of the piles or by covering static piles with a layer of composted material. The hygienic quality of the compost is further improved by running the full composting process and even more by curing the mature compost.

b) *Bioreactor-Composting.* In bioreactors, more uniform temperature profiles in the composting material can be attained, and the temperature can be influenced by the aeration rate and the technique of feeding. For a hygienically safe product, the thermophilic treatment may be of shorter duration than in the open, due to a more uniform temperature distribution. In bioreactors, the temperature must be closely monitored, and in case of disturbances, immediate countermeasures by the operator are necessary. That material which has not reached the required temperature must be used as reflux material for another passage through the reactor.

This discussion also applies to plant pathogens, especially so when composting plant residues. An exposure to 55-60°C for a few days is sufficient for the elimination of plant pathogens and nematodes. Few viruses would escape. Most plant viruses, including heat resistant ones, are destroyed by passing through the thermophilic stage of composting.

2. *End-product.* As well as controlling the process, testing of end-products can show them hygienically safe. At the time of delivery by the treatment plant to the customer, the compost should meet the following requirements:

It should not contain salmonellae (pre-enrichment) in 100 g (fresh compost);

It should not contain infective parasite ova;

It should not contain salmonellae more than 5.10, 3 faecal streptococci/g;

It should not contain salmonellae, infectious parasites such as Ascaris and Tenia, germs infectious for plants, or nematodes in an amount greater than normally found in soils.

Other groups of organisms (e.g.: enterobacteria, colibacteria, fecal-streptococci, f2-phages) are being discussed as indicators of the hygenic safety of the end-product. In some experiments, fecal-streptococci and total coliforms appear to be the best candidates as indicators, and their limit values for hygienized compost should remain, respectively, below 5.10, 3 and 5.10, 2. However, these limits may not be generalized at the moment, awaiting results of more comprehensive international experimental investigations.

Compost Classification

Products to be used in agriculture may be taken at several stages of stabilization, preceding or following the completion of processes which lead to "mature" compost. Starting from the fresh material, products should be separated according to the following classification:

Fresh Organic Matter (not to be called compost)

Definition: Raw material before undergoing decomposition (or at the beginning of it).

Agricultural uses: Unsuitable for agricultural use. However, it may serve as a substrate for composting, preparation of hot beds, preparation of mushroom compost, etc.

Hygienic evaluation: Hazardous for man, animals and plants.

General evaluation: Useful as starting material for composting. Its direct use as a soil conditioner should be discouraged because of problems leading to the diffusion of animal and plant pathogens and parasites. Special consideration should be given to the use of fresh organic matter for heat recovery (hot-bed preparation or similar uses.).

Fresh Compost

Definition: Organic matter that has been through the thermophilic stage and achieved sanitization. It has undergone partial decomposition but it is not yet stabilized into mature compost. With the exception of moisture content (which may be higher), specifications for this product must satisfy those given for "coarse compost."

Agricultural uses: This product may be used in particular crop management programs (e.g. winter top dressings in orchards, soil preparation between crops, etc.) Further decomposition and stabilization normally takes place in the soil with benefits for the improvement of soil structure and increased soil microbial activity. Maximum efficiency of stabilization is obtained by turning-in the fresh compost to a depth of 5-10 per cm, at which best oxygenation is achieved for a couple of weeks. Following this stage the material may be incorporated to full cultivation depth.

Hygienic evaluation: The thermophilic stage having been completed, the product may be considered safe for the uses proposed.

General evaluation: Use of fresh compost is possible only with proper care. Fresh compost is compatible with bulk supply and should not be packaged, as uncontrolled microbial transformations would occur.

"Compost"

Definition: The term "Compost" is a synonym of mature compost. To be thus designated, a compost must meet all of the specifications previously defined.

Agricultural uses: The material is fully suitable for application on the land, even in the presence of standing crops. Nevertheless, direct contact with the roots should be avoided (as in the preparation of potting mediums) because this may temporarily arrest growth.

Hygienic evaluations: Safe as described.

General evaluation: A valuable product for general purposes and as an organic soil conditioner on a large scale.

Cured Compost

Definition: A highly stabilized product which results from exposing "compost" to a prolonged period of humification and mineralization, beyond the stage of maturity. Minimum starting moisture content for curing should be above 35%, and should be raised to this value if necessary.

Agricultural uses: Cured compost is a highly valuable material for the preparation of artificial substrates that have direct contact with root systems of transplanted crops. Its usefulness in the nursery, flower cultivation and intensive cultivation may be matched only by that of natural humus.

Hygienic evaluation: Safe for the agricultural uses given in this paper.

General evaluation: Very valuable product for advanced uses in which it may replace peat or natural humus.

Dr. Franco Zucconi is a member of the Faculty of Agriculture at the University of Naples. Dr. Marco de Bertoldi is on the Faculty of Agriculture at the University of Udine, Italy. This paper presents a condensed form of the original document in the Proceedings of the International Symposium on Compost: Production, Quality and Use, held in Udine, Italy in April, 1986. The Proceedings are edited by M. de Bertoldi, P. Ferranti, P. l'Hermite and F. Zucconi.

SECTION FOUR

HEALTH ISSUES AND COMPOST SAFETY

15

WHEN IS COMPOST "SAFE"?

Clarence Golueke

FROM *Homo Sapiens* to *Aspergillus fumigatus,* every organism has its limits. In the biologically rich world of the compost pile, the better we understand these limits, the better we are able to manage those organisms that pose potential hazards.

The agents and mechanisms that bring about the destruction of organisms pathogenic to man and animals in the compost process are much the same as those that bring about the destruction of organisms pathogenic to plants, namely, heat, competition, antibiosis, destruction of nutrients, and time. The principal difference is in the effectiveness of the mechanisms in the destruction of the respective pathogens.

Typically, temperatures reached in a well managed compost operation are in a range of 50° to 65°C. Such temperatures are well above the thermal death points of mesophilic pathogens. However, many of the fungal pathogens of plants are facultative thermophiles. The difference between thermal tolerance possessed by animal pathogens and that by plant fungal pathogens can be illustrated through a comparison of the data in Table 1 with those in Table 2.

An important point to remember with respect to temperature tolerance is that the thermal deathpoint varies with the lifecycle stage of the pathogen in question. As a rule, reproductive forms are more resistant than are the vegetative forms. However, there are exceptions. For example, the thermal deathpoint of the spores of *Phytophthora infestans* (late blight of potatoes) is 25°C, whereas that of the mycelium is 45°C.[5] Judging from the few examples in Table 1, conidia generally seem to be less resistant to high temperatures than are mycelia.

The effectiveness of antibiosis on the destruction of human pathogens in the compost process has been described by Knoll.[6,7] He designed experiments in which he could compare the survival of inoculums of pathogens sealed in glass ampoules and inserted in a pile of actively composting material with others of the same pathogens that were not sealed, but which were distributed in the same pile. The rate of destruction of the unprotected pathogens was significantly greater than that of the sealed pathogens. Of course, inhibitive factors other than antibiosis and temperature might have been at work, as for example, inter- and intra-specific competition. However, conditions were such that theoretically nutrients were not limiting.

It should be pointed out that Gaby[8] was unable to demonstrate either bactericidal or bacteriostatic action for a wide variety of gram positive and gram negative bacteria found in composted solid waste-sludge mixtures. Nevertheless, results obtained by Knoll in his experiments are indicative of the existence of the limitations of the influence exerted by temperature. Aside from the mixed evidence afforded by Knoll's results, the likelihood of a significant role being played by antibiosis is attested by the variety of organisms, especially of actinomycetes and fungi, to be found in a composting mass of material. Fi-

TABLE 1.
Thermal Death Points of Certain Fungal Plant Pathogens[a]

Organisms	Disease	Temperature (°C)
Ustilago avenae	loose oat smut	45-53 C
U. tritici	loose smut of wheat	45-48 C
U. zeae	corn smut	106 C dry state 52 C wet state
Phytophthora infestans	late blight of potatoes	45 C mycelium 25 C spores
Taphrina deformans	peach leaf curl	46 C mycelium
Sclerospora graminicola	downy mildew of maize	40 C conidia 50 C oospores 118 C dry state
Sclerotina fructigena[b]	brown rot of stone fruits	52 C
Physoderma zeae-maydis	brown spot of corn	> 80 C
Giberella zeae	seedling blight of corn	> 65 C ascospores
Glomeralla gossypii	cotton anthracnose	> 95 C dry state 51 C wet, conidia
Guignardia bidwellii	black rot of grapes	> 80 C spores
Lentinus lepideus	brown cubical rot, western yellow pine	105 C dry 60 C wet mycelium
Macrophomina phaseoli	ashy stem blight of soy beans	55 C
Sclerotina sclerotiorum	stem rot of soy beans	50 C wet, sclerotia[c]
Pyrenophora teres	net blotch, barley	45 C conidia 55 C mycelium
Septoria lycopersici	leaf spot, tomatoes	43 C spores

[a]*From* Handbook of Biological Data, *editor, W.S. Spector, Committee on the Handbook of Biological Data, National Academy of Sciences, W.B. Saunders Co., Philadelphia, 1956 (584 pp).*
[b]*three other species—50 to 60°C*
[c]*other species—40 to 50°C*

nally, attention is called to another factor, namely the physical and chemical changes brought about in the composting mass through the combined activities of all organisms present in the mass. Good examples are the lowering of the pH level by acid formers and the elevation of the level through the destruction of acids by other microorganisms.

Competition for nutrients could be an important factor when the concentration of nitrogen or of some other essential element is in short supply. In such a situation, indigenous populations would be in the better competitive position than would be pathogenic organisms. By the very fact of being native to the material, indigenous populations would be expected to be better adapted and to be present in greater numbers than would be the pathogens which are introduced by way of contamination. Of course, competition is not necessarily limited to that for nutrients. Competition for attachment sites may be in the form of passive displacement through sheer numbers.

Destruction of nutrient derives its importance from the relatively strict nutrient requirements of pathogens in general. Pathogens generally are adapted to one nutrient source, namely, the tissue of the host. Any such material introduced into a pile by way of contamination is rapidly depleted or destroyed. Without a continuing supply of assimilable nutrient material, a pathogen can not reproduce, and will not survive long. The only recourse for the pathogen is to pass into its resistant and inactive form, if it can do so. Because their nutritional source is some form of vegetable matter, the usual major constituent of wastes destined to be composted, plant pathogens might have an advantage over human and animal pathogens.

It is convenient to treat time as a factor, too. Pathogens do not die off instantaneously—some time interval is required, however brief it may be. Actually, time provides the amalgamation or integration of all factors. Thus, it takes time for antibiosis to develop and exert its effect. It takes time for unfavorable conditions to exert an inhibitory effect on the reproduction of cells in a microbial population such that the population is eventually depleted. A negative balance between favorable and unfavorable conditions leads to a gradual destruction, the rate of which is a function of the degree of the imbalance. Finally, time has an important bearing on the safety of the use of the compost product.

Although the five mechanisms have some part in the destruction of every group of pathogens and parasites, major differences do exist between groups with respect to their responses to individual mechanisms. Thermal kill points, for instance, vary between pathogenic organisms, with fungal pathogens of plants in general surviving at higher temperatures than can be tolerated by most animal pathogens. Enteric viruses are more heat sensitive than are enteric bacteria and parasites.

TABLE 2.

Thermal Death Points of Certain Disease-Causing Organisms in Man[a]

Organisms	Temperature (°C)
Salmonella typhosa	Growth ceases at 46 C; death, 30 min. at 55-60 C
Salmonella spp	Death, 15-20 min at 60 C; 1 hr at 55 C
Escherichia coli	Death, 15-20 min at 60 C; 1 hr at 55 C
Endamoeba hisotlytica	Death, 68 C
Taenia saginata	Death, 5 min at 71 C
Trichinella spiralis	Infectivity reduces as result of 1 hr exposure at 50 C; death, 62-72 C
Necator americanus	Death, 50 min at 45 C
Brucella abortus or *sus*	Death, 3 min at 61 C
Micrococcus pyogens var. aureus	Death, 10 min at 50 C
Streptococcus pyogenes	Death, 10 min at 54 C
Mycobacterium tuberculosis var. *hominis*	Death, 15-20 min at 66 C
Mycobacterium diphtheriae	Death, 45 min at 55 C
Shigella spp	Death, 1 hr at 55 C

[a]*Golueke, C.G.* Composting: A Study of the Process and Its Principles, *The J.G. Press, Inc. Emmaus, Pennsylvania 18049, 1972 (110 pp).*

Safety of Using the Compost Product

The relation between populations of pathogens in the uncomposted waste (i.e. raw material) and the extent of the destruction of those populations during the compost process has a direct bearing on the safety of the finished compost, especially in food production. Two considerations are involved: (a) the presence of disease-causing organisms in the waste and (b) their response to the agents and mechanisms of destruction.

By their very nature, certain wastes contain human or animal pathogens; others are likely to contain plant pathogens. Examples of the former are human and animal excreta. Crop residues are examples of the latter. As a generality, it could be stated that some danger to susceptible hosts attends the use of a product produced by composting a contaminated waste, i.e., contaminated with organisms pathogenic to the crop to be cultivated or to the consumer of the crop. This generality holds unless special measures are taken to ensure the destruction of the objectionable organisms. The measures are discussed in later sections.

As for the extent of destruction accomplished in a compost operation, it depends upon the extent and modes of application of the destructive agents and mechanisms. The latter, in turn, are largely functions of the method of composting, conditions that prevail in the process, and of the care followed before, during, and after applying the process.

Human Pathogens

Sources of human pathogens in material to be composted are human body wastes, animal manures, municipal solid wastes, and plant debris. While it is true that plant debris may be a source of human pathogens, it is only an indirect source in that its pathogen content comes solely through contamination with human excreta. Human body wastes can be in the form of sewage sludge, of septic tank pumpings, and of "night soil." Pathogens in body wastes may include parasites through bacteria, viruses, and viroids. Examples of such parasites and of bacterial and viral pathogens are given in Tables 2 and 3. As for pathogenic fungi, de Bertoldi[4] names the following: *Aspergillus fumigatus, Geotrichum candidum, Trichophyton sp., Epidermophyton sp., Candida albicans, C. krusei, C. tropicalis, C. quilliermandii, Cryptococcus neoformans,* and *Trichosporon cutaneum.* The concentration of a given pathogen in a human or animal waste intended for composting is a function not only of the concentrations of pathogens in the host, but also of the age of the waste and of the treatment to which the waste has been subjected. For example, if the waste becomes a part of sewage sludge through sewage treatment, concentration of pathogens in the raw sludge would be much higher than in sludge after it has undergone anaerobic digestion.

Another source of pathogens is municipal solid waste (excluding sludge and night soil).[9] However, the concentrations of pathogens, but not necessarily of associated enteric indicator microorganisms, are much lower than those in raw sludge. Municipal solid wastes, at least in the U.S. have substantial concentrations of enteric indicators because of the large quantities of pet (dog, cat) feces disposed of with household refuse.[9,10,11]

As far as animal manures are concerned, parasites and pathogenic protozoa constitute the principal types of pathogens in them. Diesch *et al*[12,13] have done a significant amount of research on the incidence of organisms in manure that are potentially pathogenic to man. Ellis and McCalla[14] name *Salmonella, Mycobacterium, Erysepa-*

TABLE 3.

Survival Times of Animal Pathogens in the Soil and on Plants[a]

Organisms	Medium	Survival Time (Days)
Ascaris ova	soil	up to 7 years
	vegetables	27-35
Salmonella typhosa	soil	29-70
	vegetables	31
Cholera vibrio	spinach, lettuce	22-23
	nonacid vegetables	2
Endamoeba histolytica	soil	8
	vegetables	3
Coliforms	grass	14
	tomatoes	35
Hookworm larvae	soil	6 weeks
Leptospira	soil	15-43
Polio virus	polluted water	20
Salmonella typhosa	radishes	53
	soil	74
Shigella	tomatoes	2-7
Tubercle bacillii	soil	6 months
Typhoid bacillii	soil	7-40

[a]*Golueke, C. G.* Biological Reclamation of Solid Wastes, *The J.G. Press, Inc. Emmaus, Pennsylvania 18049, 1977 (249 pp).*

lothrix, Leptospira, Clostridium, and *Bacillus anthracis* as being organisms in manure that are pathogenic to man.

Control and Destruction of Pathogens

Most of the principles and practices that apply in the control and destruction of human and animal pathogens can also be applied to the elimination of plant pathogens. As one might suspect, the control and destruction of pathogenic organisms in the compost process is accomplished by composting the contaminated waste in a manner such that all pathogens are exposed to the five agents and mechanisms of destruction described earlier. How well the pathogens get exposed to the various mechanisms of destruction is a function of the method of management. Effectiveness of the exposure is measured by the extent of pathogen survival. A factor that has a key bearing on extent of pathogen destruction is the resistance of pathogenic and parasitic organisms to adverse conditions, e.g., to high temperature, low pH. For example, thermophilic organisms can withstand elevated temperatures that would be lethal to mesophilic pathogens.

As stated earlier, an organism's resistance is often a function of its life cycle stage. Animal parasites are far less resistant to destructive factors when they are in their "growth" stages than when in their resistant forms. Bacterial resistance differs with the genus and species of the bacteria—some genera and species are more resistant to adverse conditions than others. For example, *Salmonella spp* are readily destroyed through desiccation and exposure to sunlight and elevated temperatures.[15] However, they do survive cold temperatures and humid conditions. Less is known about the survival of viruses in the composting process, although experience thus far indicates a high rate of destruction.[16]

Although one would expect that pathogenic fungi could survive the compost process, de Bertoldi[4] states that all excepting *A. fumigatus* are killed in thermophilic composting. Fortunately, most pathogenic fungi either are only rarely encountered, or are present in relatively small numbers, or are pathogenic only to the unusually susceptible individual or to one whose resistance has been lowered because of poor health. *A. fumigatus* is an example of the latter.[17] Its incidence has attracted a considerable amount of attention, and consequently much has been written about the potential hazard associated with its occurrence in composting. However, most of the reports are inconclusive.[17] More recent work indicates that concentrations of *A. fumigatus* become significantly high only when woodchips or sawdust are used as the bulking agent.

Management

In this section, "Management" is interpreted as "system of composting." There are three main classes of composting systems, namely, windrow, static or forced aeration, and mechanical. Strictly speaking, the static or forced aeration system is a subdivision of the windrow class. However, it is treated as a separate division in this presentation because of the drastic difference between the conventional windrow system and the forced aeration-windrow system. Two key points are common to the three classes of systems: 1) They are designed to promote optimum aeration; and 2) They are designed to ensure the attainment of the thermophilic temperatures. The systems differ in the degree in which their design objectives are met.

For any system to be satisfactory with respect to pathogen destruction, two elements are essential: 1) All of the material to be composted, and thereby all pathogens, must be exposed to lethal conditions either simultaneously or successively. If the exposure is successive, the frequency of exposure must be such that pathogens surviving in non-lethal portions of the composting mass eventually die off as a result of repeated transfers to lethal portions, i.e., recontamination is reduced to zero. 2) Sufficient time must be allowed for the lethal factors and mechanisms to exert their full effects.

The description of the systems is kept very brief, and is confined to features that have a bearing on the destruction of pathogens.

Windrow: The windrow system, as its name implies, involves stacking the material in a long windrow pile, and of course, establishing the environmental complex required for successful composting.

Conventional windrowing differs from static or forced aeration in that aeration is accomplished by turning, i.e., tearing down and reconstituting the windrow. Of course, aeration is but one of the objectives of turning—promotion of uniform decomposition is a second; and as becomes evident later, destruction of pathogens is third. Ideally, turning would be such that material that constituted the outer layers of the pile before turning, become the innermost layers of the reconstituted pile. This ideal rarely is met, especially if turning is done by machine. In mechanical mixing, all layers in a windrow are simultaneously intermixed and agitated. The usual procedure is to have the machine start head-on at one end of the windrow and then continue through the length of the pile. The result of the simultaneous intermixing of all layers is a recontamination of "pasteurized" material with organisms that survived in the cooler layers.

Recontamination — Recontamination simply means the exposure of material pasteurized or sterilized as a result of having been in the interior of a windrow to surviving pathogens found in the outer and cooler layers. The temperature zonation in a windrow is such that a gradation from thermophilic (lethal) through mesophilic exists even during the time that biological activity is at its peak level. The depth of the mesophilic zone is a function of the heat transmission characteristics of the mixture of material, of moisture content (actually, a factor in heat transmission), and of the intensity of biological activity. (In his experience, this author has observed piles in which thermophilic zone extended to within 5 cm of the exterior of the piles.) The important fact is that pathogens can survive in the cooler zones. Therefore, when the pile is turned, the surviving pathogens are redistributed throughout the pile.

Recontamination can be counteracted by increasing the frequency of turning to a point at which surviving populations of pathogens are not allowed time to reproduce at the level required for maintaining their number. Generally, the required frequency exceeds that needed to maintain sufficient aeration.

Our work and that of others demonstrate that time can substitute for frequency—even with turning only two or three times each week a continuing decline in number of enteric organisms is observed to have taken place.[16]

Attempts have been made to set up standards for judging pathogen kill in a windrow operation and hence, safety of use of the product, that are based on the attainment and maintenance of high temperatures. For example, the designation "significant reduction" (*sic* of pathogens) implies a temperature of 50°C for 5 days, and of higher than 55°C for four hours during those 5 days. "Further reduction" implies the attainment of 55°C for 3 days in a forced aerated pile and for 5 days in a turned windrow and that the latter be turned 5 times.[18] Ware[19] suggests that a 7-day period at 55°C to 60°C be required. When human wastes are involved, governmental agencies (U.S.A.) tend to impose regulations based on time. The time is based on the number of months that have elapsed since the compost was produced. In California, the interval is as long as a year if raw sewage sludge or excreta from on-site waste treatment are involved. Of course, the time restriction does not apply if the material is heat sterilized.

Static-Pile Aeration: In forced aeration, the material is windrowed on top of a grid of perforated air ducts through which air may either be forced up through the composting mass, or be drawn through the mass and into the ducts. An important feature of the forced air approach is the capping of the material to be composted with a layer of stable material (e.g., aged compost, sawdust, leaves). The layer has an insulating effect, and if thick enough (about 30 cm), ensures the continuation of the high temperature zone throughout the material to be composted. With respect to pathogen kill, the advantage of the external layer and of the removal of the need to turn the pile are an absence of recontamination. Unfortunately, in practice, this absence is only theoretical because of the difficulty of ensuring a uniform and adequate degree of aeration throughout the composting mass. For example, short-circuiting of air may occur, or the pile may contain pockets of material in which the moisture content is less or greater than optimum. In many operations, these latter possibilities are lessened by turning the material towards the end of the compost cycle, or by storing the material in windrows for a month or longer.

With respect to standards, the regulations specify that the attainment and maintenance of a temperature higher than 55°C over a 3-day period should be sufficient to eliminate all pathogens.[18]

Mechanical: The term "mechanical" or "mechanized" applies to all systems in which composting takes place in an especially designed and closed vessel or reactor. Theoretically, the reactor is designed such that all environmental factors are maintained at as close a level to optimum as is technologically and economically feasible. The degree of closure or containment varies according to the design of the reactor and the method of operation.

The number of compost systems on the market is fairly large and is constantly changing. However, certain general designs are basic. Thus, some reactors are based on the concept of a slowly revolving drum where aeration is accomplished by tumbling. Certain systems are centered upon a vertical cylinder partitioned into two or more floors, and the composting material is dropped from floor to floor. Another version of the vertical cylinder approach is characterized by the absence of a division into floors. The use of a large rotating arm and a method of air injection through ports makes it possible to develop a well-mixed and thoroughly aerated compost. Another type of reactor seems to be a modification of the windrow system. It consists of a trough equipped with tracks on which an especially designed turning device travels. The device is a kind of a conveyor belt on wheels. Despite the protests to the contrary, composting with the use of mechanical reactor is more expensive than with a windrow system. Overall compost time, i.e., active plus maturation, is about the same as with the windrow type.

The relation to pathogen kill in a mechanical digester is a function of the type of reactor and its operations. The lethal factors and mechanisms do not differ substantially from those that occur in a windrow or in forced aeration system. Theoretically, the chances of a complete destruction of pathogens are greater in a mechanized, enclosed reactor than in a windrow system because lethal temperatures can be made to pervade the composting mass. In actual practice, however, pockets may occur in which the temperature does not reach the thermal deathpoint. With the large masses of materials involved, it seems probable that at least some isolated particles escape exposure to lethal conditions. On the other hand, enclosed systems offer the advantages that attend a close control.

If lethal temperatures could be assumed to prevail throughout the reactor, then the exposure time required for pathogen kill would become a matter of hours rather than days.

Safe Use of the Product

Even when all pathogens are assumed to have been killed in the composting process, prudence dictates a certain amount of caution in the use of the compost product. Complete destruction simply cannot be unequivocally guaranteed. Given this qualification, it can be stated that the identity of the pathogens and the probable extent of their survival should determine the applications for which the product can be safely used. As a generalization, it may be stated that the possibility of pathogens being present in the raw wastes and in the compost product is quite remote if no body (human) wastes and wastes from diseased animal are involved.

If human wastes and those from diseased animals are involved, and pathogens in them do survive the compost process, their survival under field conditions becomes another consideration. Much has been reported about such survival. Examples of survival times in the soil and on plants are given in Table 3. According to the data in the table, survival times are as varied as the types of pathogens. For instance, *Endomoeba histolytica* survives only 3 days in soil and 8 days on vegetables, whereas *Ascaris* ova can live as long as 27 to 35 days on vegetables and as long as 7 years in the soil. Dr. Sagik[19] reports the isolation of poliovirus II and Coxsackie B-3 from the soil more than 6 months after the injection of sludges into the soil. He states that sludge-associated viruses may be retained effectively within the sludge-soil matrix.

The type of crop for which the compost is to be used determines the extent of pathogen destruction required. If sewage sludge or other forms of body wastes are composted, then complete destruction of pathogens is essential before the compost product can be safely used for root crop and leafy vegetable production, that is, if the vegetables are destined to be eaten raw. Unless a span of at least a year is involved, such compost should be pasteurized. If the vegetables are to be cooked, then the requirement is not as rigorous. Requirements for use in orchard production are not as strict as for root crops and leafy vegetables. The major problem is the possibility of fallen fruit becoming contaminated. Use in the cultivation of grains, cereals, and in silviculture is even less demanding in terms of destruction.

Plant Pathogens

Before proceeding with the discussion on plant pathogens, it is emphasized that the matter of the inhibitive effect exerted by immature compost is not covered. It has been well treated by Zucconi[12], and by DelVleeschauwer, *et al.*[3]

The principal source of plant pathogens in a composting operation are the wastes from food preparation and the discards from food markets. However, a number of factors determine whether diseased debris and wastes can serve as a source. One is plant specificity. Here, pathogenicity would depend upon contact with a susceptible plant. A second factor is the variation in degree of specificity that exists among groups of pathogens, in that some type of plant pathogens can infect a wider variety of plant species and genera than can others. The specificity also varies with the growth stage of the plant. Vigor of a plant is another important factor. This latter brings up the interesting possibility that the increase in vigor and growth of the crop plants brought about by the application of a contaminated compost may be great enough to prevent or to mask inroads by the pathogens in the compost.

Types of Plant Pathogens

As with human and animal pathogens, types of plant pathogens include parasites, bacteria, viruses, and fungi. As long ago as 1966, P. Martin[20] in a one-page paper summarized the subject of the bearing of composting on plant pathology. In his summary, he briefly described four of the principal species that are pathogenic to plants, namely, *Plasmodiophora brassicae, Olphidium brassicae, Rhizoctonia solani,* and *Heterodera rostochiensis.* A more extensive list of plant pathogens may be found in Tables 4 and 5. Table 4 is a list of pathogenic bacteria and of the diseases caused by them. Examples of plant viruses and their respective diseases are given in Table 5. With respect to pathogenic bacteria, the fact of plant specificity must be especially kept in mind.

Viruses constitute a serious problem in plant pathology, and hence require special attention in composting and in the use of the compost product. In terms of pathogenicity, incidence, and diversity, fungi are more significant to plants than to humans and animals. In addition to the lower forms of pathogens are a host of higher forms of

TABLE 4.

Examples of Bacteria Pathogenic Pathogenic to Plants

Organisms	Plant: Scientific Name	Plant: Common Name	Disease
Erwinia amylovora	Pyrus malus	apple	fireblight
Pseduomonas coronafaciens	Hordeum vulgare	barley	leafspot
Corynebacterium flaccunifaciens	Phaseolus vulgaris	vulgaris bean	wilt
Pseudomonas wieringae	Beta vulgaris	sugar beet	ring rot
Xanthomonas campestris	Brassica oleacea var. capitata	cabbage	black rot
Erwinia carotovora	Daucus carota	carrot	soft rot
Pseudomonas apii	Apium graveolens	celery	leaf spot
P. radiciperda	Trifolium spp	clover	root rot
Bacterium stewartii	Zea mays	corn	wilt
X. malvacearum	Gossypium spp	cotton	angular leaf blight
X. campestris	Crucifera	cabbage, cauliflower, turnip	black rot
E. tracheiphia	Cucurbits		wilt
Agrobacterium tumefaciens		fruit trees	crown gall
E. vitivora	Vitis vinifera	grapes	dead arm
P. allicola	Alium cepa	onion	scale rot
P. pisi	Pisa sativum	peas	blight
X. pruni	Prunus persica	peach	spot leaf or fruit
Erwinia spp	Solanum tuberosum	potato	soft rot
Solanaceae		eggplants, pepper, tomato	wilt
P. glycenea	Glycine max	soybeans	leaf spot
C. michiganense	Lycopersicum esculentum	tomato	canker and wilt

Generalization: Erwinaceae—amylovora grup = necrosis; caotovora = soft rot

life (e.g., nematodes, adult and larval forms of insects). On the other hand, the number of higher plants that are parasitic (e.g., dodder, mistletoe) is relatively small.

The fundamental principles of control and destruction in the compost process are comparable to those of human and animal pathogens. However, much remains to be learned about the extent and survival of plant pathogens, especially of the viruses. A problem that can complicate the destruction of plant pathogens is that many plant pathogens are facultative thermophiles, especially the fungal forms. However, Hoitink *et al*[21] writes of the thermokill of *Phytophthora cinnamoni, Pythium irregulare, Rhizactonia solani, Botrytis cinerea,* and *Erwinia carotovora* var. *chrysanthemi.* Certain viruses can withstand high temperatures, e.g., tobacco mosaic.[20] On the other hand, other viruses are vulnerable to them.

The approach to effective management is identical with that described in the section on human and animal pathogens, inasmuch as the principles involved are the same.

Use of the Product

The relation of the extent of pathogen destruction to the safe use of the compost product depends largely on how well pathogens that may be in the compost product can survive under field conditions. For example, Martin[20] claims that the resistant spores of *P. brassicae* can contaminate an acid soil for 4 to 6 years. (They quickly disappear in an alkaline soil.) Examples of survival times of several plant viruses outside the host plant are given in Table 5. According to the table, survival times in plant juices are far shorter than on dried tissue.

In the absence of more extensive and definitive formal studies on survival, it would be more prudent to assume survival. Therefore, to provide a safety factor, it would be advisable to avoid the use of a compost product made from diseased plant debris in the cultivation of plants of the same species as the diseased plants, or of plants susceptible to the disease.

Fungicidal Properties of Composted Bark

In the past decade or two, a very interesting development has occurred, namely, the appearance of reports on the toxicity of certain composted barks to pathogenic fungi and to nematodes.[22-25] In his paper on the fungicidal properties of composted bark, Hoitink[22] states that the partial replacement of peat by composted bark in con-

TABLE 5.

Survival Time of Plant Viruses Outside Plant[a]

Organisms	Disease	Medium	Time (days)
Marmor medicaginis	alfalfa mosaic	plant juice	8-9
M. phaseoli	bean mosaic	plant juice	2-24 hr
M. cruciferarum	cauliflower mosaic	plant juice	14-15
M. umbilliferum	celery mosaic	plant juice	6-7
M. cucumeris	cucumber mosaic	plant juice	2-5
M. terrestris	oat mosaic	dry tissue	177
Annulus dubius	potato mottle	plant juice	3-4 mo
		dried tissue	363
Ruga verrucosans	sugar beet curly top	plant juice	1-14
		dried tissue	10 mo
M. flavopunctum	bean yellow stipple	plant juice	5-6
		dried beans	80
M. angustum	false garlic mosaic	plant juice	17-52
		dried tissue	40-43
A. tobaci	tobacco ringspot	plant juice	1-6
		dried tissue	393
M. tritici	wheat mosaic	dried tissue	290

[a]*From* Handbook of Biological Data, *editor, W.S. Spector, Committee on the Handbook of Biological Data, National Academy of Sciences, W.B. Saunders Co. Philadelphia 1956 (584 pp)*

tainer media has become an added element in the control of root rot of container-produced ornamentals, because composted tree bark is inhibitory to the development of many pathogens. He traces the beginning of the recognition of the toxicity of certain composts to plant pathogens to the early 1960's. At the time, reports of observations made in nurseries began to appear, which indicated that philodendrons produced in media amended with composted bark were less damaged by Photophthora root rot, than were those grown on media amended with peat.[26] Recognition was slow in developing because a part of the toxicity could be attributed to the ammonia released when the ammoniated tree bark compost generally used at the time began to deteriorate. Reports of further experience are to the effect that in many nurseries, philodendron root rot does not become a problem when most or all of the Canadian or German peat in container media is replaced by composted hardwood bark. Recent practice shows that root rots do not occur in media in which the organic component is a 4:1 (v/v) mixture of bark and peat.

Experience in the past decade is beginning to indicate that hardwood bark is quite effective in suppressing soil-borne disease in floricultural crops and in foliage plants. For example, Tajan[23] has demonstrated the effectiveness of composted hardwood bark in controlling nematode infestation of citrus crops.

Reported results of studies indicate that the pathogen toxicity of composted bark diminishes as the ratio of wood to bark increases. In fact, the suppressive effect on phytophthora root rot is nullified when the ratio of wood to bark reaches 6:1. The suppressive effects also is lost when the volume of peat is equal to or greater than that of bark. The species of the trees from which the bark was obtained is another factor that determines the species of pathogens susceptible to the suppressive effect of a given composted bark. In a greenhouse study, Hoitink[27] found that Phytophthora, Pythium, and Thielaviopsis root rots were suppressed when inoculated into a medium that contained compost from a mixture of hardwood species. Phytophthora and Pythium root rots were suppressed by composted pine bark, but Rhizontonia damping off was not affected.

Mechanism of Suppression: Hoitnik, *et al*[26] observed that leachates from a bark compost in which rhododendrons had been grown for two years had no inhibitive effect on otherwise susceptible pathogens; whereas, those from a hardwood bark-sand compost had such an effect. From that observation he concluded that the suppressive effect of compost bark is due to chemical and biological factors rather than to physical factors. Hoitink[29] suggests three mechanisms that may play a role in the suppression of plant pathogens in media amended with tree bark: 1) improved aeration resulting from the incorporation of the bark particles; 2) composts support high levels of antagonists and phagous organisms; and 3) the possession of chemical inhibitors. According to Hoitink, evidence is accumulating that emphasizes the importance of the third mechanism. In the same paper, he states that suppression apparently is not a result of a reduction of inoculum. For evidence, he refers to results reported by Chef.[27] According to the results, death rates of *Fusarium oxysporum* f. sp. *Chryanthemi propagules* in peat and composted bark media were not significantly different over a 100-day period. Hoitink refers to another study in which it was found that the suppression of *P. cinnanaonii* in pine bark media as compared with that in peat, was not due to differences in the death rate of chlamydospores.

References

1. Zucconi, F., A. Pera, and M. de Bertoldi, "Evaluating Toxicity of Immature Compost," *BioCycle, 22*(2):54-58 (March-April, 1981).
2. Zucconi, F., M. Forte, A. Monaco, and M. de Bertoldi, "Biological Evaluation of Compost Maturity," *BioCycle, 22*(4)27-30, (July-Aug., 1981).
3. De Vleeschauwer, D., O. Verdonek, and P. Van Asche, "Phytotoxicity of Refuse Compost," *Biocycle, 22*(1):44-46, (Jan.-Feb., 1981).
4. de Bertoldi, M. "Pathogenic Fungi Associated with Land Application of Sludge," World Health Organization, Regional Office for Europe, January, 1981.
5. *Haridbook of Biological Data,* National Academy of Sciences, W. B. Saunders Co., Philadelphia and London 1956, (584 pp).
6. Knoll, K.H., "Composting from the Hygienic Viewpoint," International Research Group on Refuse Disposal, IRGR Information Bulletin 7, p. 10, July 1959.
7. Knoll, K.H., "Influence of Various Composting Processes on Non-Sporeforming Pathogenic Bacteria," International Research Group on Refuse Disposal, IRGR Information Bull. 19(1), 1963.
8. Gaby, W.L., *Evaluation of Health Hazards Associated with Solid Waste/Sewage Sludge Mixtures,* EPA-670/2-75-023, U.S. EPA, National Environmental Research Center, Cincinnati, Ohio, 1975.
9. Ware, S.A., "A Survey of Pathogen Survival During Municipal Solid Waste and Manure Treatment Processes," EPA-600/8-80-034, Municipal Environment Research Laboratory, Office of Research and Development, U.S. EPA, Cincinnati, Ohio 45268, Aug. 1980.
10. Cooper, R.C., *et al, Effect of Disposable Diapers on the Composition of Leachate from a Landfill,* Sanitary Eng. Research Lab Report No. 74-3, University of California, Berkeley, 1974.
11. Peterson, M.L., *Parasitogical Examination of Compost,* Solid Waste Research Open File, U.S. EPA, 1971.
12. Diesch, S.L., B.S. Pomeroy, and E.R. Allerd, "Survival and Detection of Leptospires in Aerated Beef Cattle Manure," Paper presented at International Sysposium on Livestock Wastes, Ohio State University, Columbus, Ohio, March, 1971 (17 pp).
13. Diesch, S.L., B.S. Pomeroy, and E.R. Allerd, *Survival of Pathogens in Animal Manures,* Final Report, College of Vetinary Medicine, University of Minnesota, St. Paul, Minn., 1971 (189 pp).
14. Ellis, J.R. and T.M. McCalla, "Fate of Pathogens in Soils Receiving Animal Wastes—A Review," *Transactions of the A.S.A.E., 21*(2):309-319 (March-April, 1978).
15. Hales, D., "Salmonellae in Dried Sewage Sludge," *Environmental Health (Journ. of the Assoc. Health Inspectors, Gr. Br.),* 82(11):213-215 (Nov. 1974).
16. Cooper, R.C. and C.G. Golueke, "Survival of Enteric Bacteria and Viruses in Compost and its Leachate," *Compost Science/Land Utilization, 20*(2):29-35 (March-April, 1979).
17. Oliver, Jr. W.M., "The Life and Times of *Aspergillus fumigatius,*" *Compost Science/Land Utilization, 20*(2):36-39 (March-April, 1979).
18. *Technical Bulletin: Composting Processes to Stabilize and Disinfect Municipal Sewage Sludges,* EPA-430/9-8-011, U.S. EPA Office of Water Program Operations, Washington, D.C. 20460, July 1981.
19. Sagik, B.P., *Virus Survival in Soils Injected with Municipal Wastewater Treatment Residuals,* Report No. 78-7, Executive Summary of Final Report submitted to Advanced Environmental Research and Technology, NSF-RANN, Washington, D.C. 20550, June 1978.
20. Martin, P., "Plant Pathology Problems in Composting," *Compost Science, 6*(3):23 (Autumn-Winter, 1966).
21. Hoitink, H.A.J., L.J. Herr, and A.F. Schmithenner, "Survival of Some Plant Pathogens During Composting of Hardwood Tree Bark," *Phytopothology, 66:*1369-1372, 1976.
22. Hoitink, H.A.J., "Fungicidal Properties of Composted Bark," *Compost Science/Land Utilization* (Nov.-Dec. 1980).
23. Tajan, A.C., "Use of Municipal Refuse Compost on Nematode Infested Citrus," *Citrus and Vegetable Magazine, 40*(6):44-49 (Feb. 1977).
24. Hoitink, H.A.J. and H.A. Poole, "Bark Compost Use in Container Media," *Compost Science/Land Utilization, 21*(3):38-41 (May-June, 1980).
25. Malek, R.B. and J.B. Gartner, "Hardwood Bark as a Soil Amendment for Suppression of Plant Parasitic Nematodes in Container Grown Plants," *Hort. Sci., 10:*33-35, (1975).
26. Hoitink, H.A.J., D.M. Van Doren, Jr., and A.F. Schmithennex, "Suppression of Phytophtora cinnamonia in a Composted Hardwood Bark Potting Medium," *Phytopathology, 67*(4):561-565 (Apr. 1977).
27. Chef, D.G., *Suppression orium Wilt of Chrosanthemum in Composted Hardwood Bark,* M.S. thesis, Ohio State University, Columbus, Ohio, 1977.

16

CRITERIA FOR USE OF SLUDGE COMPOSTS

REGULATORY requirements for the marketing and distribution of sludge compost were the subject of a joint study of the Water Quality Branch of EPA Region I and the Massachusetts Department of Environmental Quality Engineering. A primary purpose of the 1986 study was to determine the feasibility of end product use in connection with Boston's long-term sludge management program. A related purpose was to assist states within the region to assess regulations.

To help develop recommendations for the region, Metcalf & Eddy did an exhaustive literature search to evaluate the scientific basis for regulation of metals, organic chemicals, and pathogens in composted sludges applied to land. The firm also assessed the soil chemistry and site conditions typical of the Northeast which can play a limiting role in the land application of sludges. One result was a section of the report that included conclusions and recommendations about acceptable limits for inorganic and organic contaminants of sludge, reasonable annual and cumulative maximum application rates, and acceptable levels of stabilization. The conclusions are presented in three broad groups: inorganic contaminants (cadmium and lead), organic chemical contaminants, and pathogenic organisms. The following information is adapted from the final report titled *"Sludge Compost Marketing and Distribution Regulatory Requirements In The United States."*

The conclusions encompass acceptable limits and application rates for specific constituents, but do not address site selection and land management practices except in general terms. The recommendations focus on factors that most influence human rather than animal exposures to pathogenic constituents in sludges.

Cadmium: The report recommends that the maximum allowable concentration for cadmium in sludges or sludge products to be applied to land should be 13 mg/kg dry weight. Although several investigators and review papers commonly endorse 25 mg/kg DW cadmium as an appropriate maximum level in "domestic" sludges,[1,2,3] the scientific basis underlying this number was never clearly stated.

To substantiate this recommendation, the report states the following:

"Some of the most convincing recent evidence supporting the safety of low metal sludges comes from work by Chaney et al.[4] which studied cadmium uptake in lettuce leaves from plots receiving sludge. Lettuce has a strong tendency to accumulate cadmium as do several other plants such as spinach, chard, turnip and beet greens, and carrots.[1] Plots receiving sludges with 13.4 mg/kg cadmium showed no demonstrable increase in lettuce leaf cadmium concentrations at application rates up to 3 kg cadmium/ha. The difference in lettuce leaf cadmium levels between lettuce grown on soils at pH 5.6 and 6.4, was negligible indicating that the bioavailability of cadmium is restricted even at the low pH levels more typical of northeast soils. Chaney et al.[4] and Logan and Chaney[5] suggest but do not state that the bioavailability of cad-

TABLE 1.

Recommendations For Cadmium And Controlled Application

Regulations	Cadmium	Lead	PCBs
Acceptable Limits	25 mg/kg DW	500 mg/kg DW	10 mg/kg injected 3 mg/kg surface
Application Rates	a. Use federal rates for cadmium to establish upper limit. b. Incorporate strong recommendations that sludge be used only at agronomic rates 10-20 Mt/ha for most agricultural uses.		
Maximum Cumulative Concentrations	Same as current federal regulations	500 kg/ha	
Application Restrictions			
1. General	Never to growing crops—for direct human consumption (i.e., only to soil). Injection or incorporation recommended even for forage crops to be harvested for animals intended for human consumption.		
2. Crops	Never for growing tobacco	—	—
3. Grazing Restrictions	a. Three months minimum should be required between application and grazing by animals intended for human consumption. Otherwise, one month or b. One month minimum. Use of animal products (meat or dairy) restricted for 3 to 8 months from onset of grazing.		
Background Metal Monitoring at Soil Required	Yes a. Annual basis of sludge applied more than once/year. b. Before application otherwise.	Yes	—

mium at concentrations of 25 mg/kg would be similarly low.

"Chaney's work is particularly important because it appears that previous arguments for lower maximum acceptable levels of cadmium in sludge were based on studies using cadmium salts added to soils. Such studies predicted a linear increase in plant cadmium content with increasing cumulative application rate regardless of initial sludge concentration. Chaney's work with cadmium applied in sludges indicates that this relationship does not hold. Other investigators suggest that chemical factors in the sludge, expressed generally as the "sludge-specific cadmium adsorption capacity," limit cadmium's bioavailability.[5] Only at high cadmium levels—levels beyond 25 mg/kg—is the sorption capacity of the sludge apparently exceeded. The initial concentration of cadmium is subsequently a strong predictor of available cadmium for long periods beyond this initial sludge application."

Lead: The basis for a maximum acceptable level of lead in sludges is more uncertain, say the authors of the Metcalf & Eddy report. "The maximum level of lead recommended by most studies is 100 mg/kg although the public health basis for that number was never explained. The USEPA, USFDA, and USDA[1] recommended a maximum final concentration of lead in soils to be 800 mg/kg. Again, the public health basis for their recommendation was not given.

"This report recommends that 500 mg/kg DW lead be the maximum acceptable level for lead in sludges or sludge compost destined for sale/give-away programs. This recommendation is based on the assumption that children could ingest small amounts of sludge containing

TABLE 2.

Recommendations For Pathogens In Composted Sludge: General Public Use

Regulation or Guideline	Recommendation
Degree of Stabilization	PFRP
Labeling Requirements	1. Recommended sludge application rate based on soil conditioning requirements 10-20 Mt/ha - 50-100 lbs/250 sq. ft. 2. Recommend immediate incorporation into soil. 3. Recommend washing of all fruits and vegetables grown on sludge-amended soils.

TABLE 3.

Recommendations For Pathogens In Composted Sludge: Controlled Land Application

Regulations or Guidelines	Recommendation
Degree of Stabilization	PSRP
Application Rates	1. Establish upper limit on the basis of maximum annual loading rate for cadmium or other limiting contaminant. 2. Incorporating strong recommendations that sludge be used only at agronomic rates, 10-20 Mt/ha, for most agricultural uses.
Application Restrictions	
1. General	Never directly to crops grown for direct human consumption.
2. Between Hay Crops	Apply within 1 week of first harvest to minimize sludge adherence to vegetation. One month minimum required between hay crops.
Grazing Restrictions	Establish a waiting period of 2-3 months after application.

lead at these levels and based on the studies cited by Logan and Chaney,[5] could therefore, take in unacceptably high levels of lead. In reality, sludges used in the home are likely to be mixed in garden soils or potting soil thereby reducing the overall level of lead in soils that a child may have access to. This factor appears to add an additional margin of safety to the 500 mg/kg lead recommended. However, there was no opportunity to review the studies from which the 500 to 1000 mg/kg values were derived for this study, so definitive support for the applicability of the values to sludge regulations cannot be established at this time.''

Organic Chemicals: The report states that insufficient data were available to evaluate the public health basis for limits on organic chemicals in municipal sludges intended for general public use. It adds that sludges packaged or otherwise designated for this use should clearly state the maximum metal and organic chemical concentrations permitted and those detected in the sludge so that individuals know the characteristics of the material they choose to use.

Tables 1, 2, and 3, excerpted from the Metcalf & Eddy report, summarize other recommendations with regards to compost distribution and use. In their conclusion, the authors acknowledge that the public health and environmental impact of land application of sludges is the subject of ongoing debate. They state that if sufficient exposure to the various toxic metals, organic compounds, and pathogenic organisms occurs, there is a potential hazard to public health and the environment. ''The current debate focuses on the most effective means of reducing the likelihood of exposure and impact. The consensus emerging from this review of recent scientific literature is that a combination of scientifically based limits on toxic and pathogenic constituents of sludge and sound land management practices can effectively protect both public health and the environment.''

References

1. U.S. EPA, U.S. FDA, USDA 1981. *Land Application of Municipal Sewage Sludge for the Production of Fruits and Vegetables; A Statement of Federal Policy and Guidance.*
2. Hornick, S.B., L.J. Sikora, S.B. Sterrett et al., 1984. *Utilization of Sewage Sludge Compost as a Soil Conditioner and Fertilizer for Plant Growth.* U.S. Dept. of Agriculture, Agriculture Information Bulletin No. 464.
3. Pennsylvania State University, 1985. *Criteria and Recommendations for Land Application of Sludges in the Northeast.* Bulletin 851, March.
4. Chaney, R.L., S.B. Sterrett, M.C. Morella, and C.A. Lloyd, 1982. ''Effect of Sludge Quality and Rate, Soil pH and Time on Heavy Metal Residues in Leafy Vegetables.'' In: *Proceedings of the Fifth Annual Madison Conference on Applied Research and Practice on Municipal and Industrial Waste,* Univ. of Wisconsin, Madison.
5. Logan, T.L. and R.L. Chaney, 1983. ''Utilization of Municipal Wastewater and Sludge on Land-Metals.'' In: *Proceedings of the 1983 Workshop on Utilization of Municipal Wastewater and Sludge on Land.*

17

DETERMINING PATHOGEN LEVELS IN SLUDGE PRODUCTS

Nora Goldstein, William A. Yanko, John M. Walker, Walter Jakubowski

A YEAR long search for pathogens in sludge products that could be distributed and marketed (D&M) to the general public was completed in 1988 for the U.S. Environmental Protection Agency by the County Sanitation Districts of Los Angeles County. The study analyzed 498 samples of sludge products that were taken serially from 26 locations in 17 states over a period of one year. Each sample was analyzed for over 15 pathogen and indicator microorganism groups. The study found that the sludge products were remarkably free of pathogenic viruses and viable parasitic ova.

The study did find, however, that the ubiquitous pathogenic bacterium *Salmonella* was detected in a significant number of the samples. *Yersinia* was found in samples of sludges from a far fewer number of the locations. As might be expected, indicator microorganisms were found in all samples. The occurrence of *Salmonella* and *Yersinia* in D&M products may represent a possible health problem under certain scenarios. However, the study was designed only to determine the occurrence of pathogens in the D&M products. The possible health effects to users of the products was not within the scope of the study and no conclusions could be developed on actual health risks.

Furthermore, the exact significance of the presence of *Salmonella* is unclear because this pathogen occurs commonly in the environment, and has the capability to regrow in previously "clean" substrate such as disinfected sludge products and food. The significance of the *Yersinia* occurrence also is unclear. Some species of this bacterium are pathogenic and some are not. Further analysis of a small sample of the *Yersinia* species isolated in this study did not reveal them to be pathogenic.

Review of the data gathered in this recent investigation is very encouraging for compost facilities. While *Salmonella* was detected, one of the facilities studied has found that through modified composting practices (explained later in this chapter), it can be controlled.

The San Jose Creek Water Quality Laboratory (part of the Los Angeles County Sanitation Districts) in Whittier, California, conducted the study under contract with EPA's Health Effects Research Laboratory in Cincinnati, Ohio. The basic objective of the search was to determine the types and concentrations of various non-pathogenic indicator microorganisms and pathogenic microorganisms that may be present in distributed and marketed sludges. The findings were published in a report entitled: "Occurrence of Pathogens In Distribution And Marketing Municipal Sludges" (NTIS: PB88-154-273/AS). The study also surveyed the presence of chemicals in distributed and marketed sludges. Those results appear in a separate report, "Trace Organics and Inorganics In Distribution & Marketing Municipal Sludges" (NTIS: PB88-160-585).

Sludge Product Sampling

Composts and other sludge products were sampled from a total of 26 facilities (see Tables 1 and 2). Composts (treated by processes to further reduce pathogen-PFRP)

TABLE 1.

Weekly Sampling Sites*

Location	Sampling Point
1. Pennsylvania Static Pile Composting Facility	a) Giveaway Bin b) Screened Compost
2. California Windrow Composting Facility	a) Final Sawdust Compost b) Final Recycle Compost c) Bagged Compost d) Bagged Compost with Rice Hull e) Bagged Compost with Sawdust f) Bagged Compost with Bark and Redwood

**Sampled weekly from Jan. 1986-Jan. 1987.*

from two facilities—a windrow operation and an aerated static pile operation—were sampled weekly for one year. Material, treated either by a PFRP or PSRP, from the other 24 facilities was sampled on a bimonthly basis for one year. Samples were analyzed for a variety of bacteria, viruses, parasites and fungi in the indicator and pathogen categories (see Table 3).

Five compost products were sampled weekly at the windrow facility. They included the final compost produced, and four commercially marketed, compost-based soil amendment products. The two weekly sampling points at the static pile facility were the final screened compost that was utilized in a number of bulk distribution programs and the "giveaway bin" that contained unscreened compost available to the public for home use. Operators at both facilities assured the researchers that they were following recommended guidelines for composting (i.e. meeting time/temperature requirements for a PFRP product).

Study Results

Most classes of pathogens were not detected in most of the 354 sludge compost samples taken from the two facilities tested each week. The only potential pathogens that were detected were bacterial, including *Salmonella, Yersinia,* and toxigenic *Escherichia coli.* On the other hand, the indicator microorganisms were frequently detected at high concentrations. Tremendous indicator variability was observed with some concentrations varying by as much as 10 orders of magnitude, notes the report. It states:

Data from the windrow site grouped into two strata. The first stratum included the final compost and the commercial product containing only screened final compost. Average concentrations of microorganisms were higher in

TABLE 2.

Bimonthly Sampling Sites*

Location	Process	Classification
Maine	Static Pile	PFRP
New Jersey	In-Vessel	PFRP
Maryland	Static Pile	PFRP
Virginia	Static Pile	PFRP
Virginia	Aerated Windrow	PFRP
North Carolina	Static Pile	PFRP
Florida	Drying Bed	PFRP
Florida	Aerobic Digestion Air Dried	PSRP
Georgia	Heat Drying	PFRP
Indiana	Static Pile	PFRP
Indiana	Zimpro/Filter Press	PFRP
Texas	Air Dried	PSRP
Kansas	Air Dried	PSRP
Kansas	Windrow	PFRP
Colorado	Air Dried	PSRP
Utah	Aerobic Digestion Air Dried	PSRP
Montana	Proprietary	PFRP
Colorado	Aerated Windrow	PFRP
California	Windrow	PFRP
California	Static Pile	PFRP
California	Air Dried	PSRP
Arizona	Air Dried	PSRP
Arizona	Air Dried	PSRP
Oregon	In-Vessel	PFRP

**Sampled each every two months from April 1986–April 1987.*
PFRP: Process to Further Reduce Pathogens
PSRP: Process to Significantly Reduce Pathogens
Note: Air dried sludges were anaerobically digested unless otherwise stated.

TABLE 3.

Microorganism Categories Selected for Analysis

Indicators	Pathogens
Total coliform	Enterotoxigenic *E. coli*
Fecal coliform	Total enteric bacteria
Fecal streptococci	*Salmonella*
Aerobic plate count	*Campylobacter*
Anaerobic plate count	*Yersinia*
Total Fungi	*Ascaris ova*
Thermophilic fungi	Total parasites
Bacteriophage	Total enteric viruses

the bagged compost than in the field compost samples but the differences were not significant at the 95 percent confidence limits (C.L.). The second grouping of data consisted of bagged commercial products containing additional amendments (added to the material after the compost operation); these products contained significantly higher concentrations of microorganisms than those in the first stratum.

Analysis of the data from the static pile composting facility indicated that these two sampling points were also significantly different. The screened compost contained higher levels of bacteria than the material in the giveaway bin. Some, but not all, of the difference could be accounted for by the presence of the woodchips in the unscreened compost.

Results from the 24 facilities sampled on a bimonthly basis also showed a great deal of variation in microbial densities. Air dried sludges often contained lower concentrations of enteric bacteria than composted sludges and as a group, were significantly lower than static pile composting sites. *Salmonella, Yersinia* and toxigenic *E. Coli* were randomly isolated in the various sludge products from the facilities sampled bimonthly, generally at low levels and at lower levels than in the products from the two facilities that were sampled weekly.

Conclusions

Some of the conclusions provided in the report are:

- No health hazard was found associated with respect to parasitic helminth ova, *Campylobacter* (a bacterium that can cause gastroenteritis, or enteric viruses. In fact, no campylobacters or viable parasitic helminth ova were detected in any sample and viruses were detected in only two of the 498 samples. The test for campylobacters in compost was relatively ineffective but other available data suggest these bacteria would not survive composting or air drying. Helminth ova were detected regularly but no indications of viability were observed. No protozoan cysts were found.
- All of the sewage sludge products examined were found to contain variable concentrations of nonpathogenic indicator microorganisms. Some products contained bacterial pathogens—most specifically *Salmonella*—at relatively high frequencies and levels. Variability of microorganism concentrations was often great between different facilities and between different samples from the same facility. Many of the observed trends would not have been detected without a large number of samples collected over a long period of time.
- Overall, the highest concentrations of microorganisms occurred in samples from static pile composting systems; the lowest concentrations were found in pelletized sludge from a heat drying process. Microorganism densities in aged anaerobically digested-air dried sludges were as low as, or lower, than most of the composted sludges.
- Composts from the windrow facility sampled weekly that were modified with various materials to produce commercial soil amendments contained significantly higher concentrations of bacteria and fungi than the base compost material. The data suggested a nutrient-related regrowth phenomena.
- Potentially pathogenic bacteria, including *Salmonella* sp., *Yersinia* enterocolitica and toxigenic *E. Coli* were detected. *Salmonella* sp. were the most frequent pathogen detected. The quantitative test for toxigenic *E. Coli* indicated that these strains, when present, occurred at very low levels. However, the percentage of colonies that were toxigenic strongly suggested that the concentration of toxigenic strains was much higher than indicated. *Yersinia enterocolitica* occurred at very high densities in some samples. The isolation of *Yersinia* was consistent with a seasonal occurrence, with the prevalence and density higher in colder months. Based on a small number of tests the *Yersinia* appeared to be avirulent (not causing disease).
- Given the considerable variation observed in microbial densities and the reasonably frequent isolation of *Salmonella,* bacterial monitoring to assure product quality may be of value for the home use of sludge and compost soil amendments. Regression analysis suggested that total or fecal coliforms or fecal streptococci may be suitable indicators for monitoring.

It was also concluded that the occurrence of pathogenic bacteria could present a potential health problem under certain circumstances. These is, however, a discussion in the report about how the windrow facility has been able to modify its composting practices to essentially eliminate the occurrence of *Salmonella* (see below).

Public Health Effects

The report states that the extent of risk associated with the use of sludge products has not been determined. However in conversations with California health officials and EPA representatives (as well as a literature search), the researchers found no reports of anyone ever getting sick from using a sludge compost-based fertilizer or soil amendment product. What is not known is whether or not there is any measurable risk of infection associated with using products containing the levels of *Salmonella* detected in this study.

The report included some discussion on the health significance of *Salmonella* and *Yersinia:*

"Rates of *Salmonella* infections have been increasing in the United States in recent years. Salmonellosis clearly remains a disease of concern. Nevertheless, the salmonellae are fairly ubiquitous and it is common knowledge that salmonellae may be found in food products such as chicken and turkey. The World Health Organization (1981) has suggested that it may be impossible to eliminate salmonellae from the environment and that the best control measures may be increased education about food handling practices.

"It is difficult to assess the potential health significance of the salmonellae detected in the compost products. There are no known cases of salmonellosis traceable to the use of sludge based soil amendments. The studies by Ottolenghi *et al* (1987) found no apparent risk to farm families using anaerobically digested sludges in agricultural applications. These sludges were shown to contain salmonellae. Available data suggest that salmonellae may persist in sludge amended soil for up to five months but that a 90% reduction occurs within three weeks (Sorber and Moore 1986). There is some evidence that specific *Salmonellae* serotypes may selectively grow in compost and that these strains are less commonly associated with clinical infections (unpublished data).

"All of these factors appear to mitigate the health significance of the salmonellae data. At the same time the question raised by these results should not be ignored. Additional research to better understand the reasons for the high salmonellae levels is warranted. Relatively simple management practices may significantly alleviate any potential hazards associated with salmonellae in compost."

To further understand the health significance of *Yersinia,* the San Jose Creek Water Quality Laboratory contracted with the New York State Department of Health to do a characterization of some *Yersinia* isolates. The isolates were non-pathogenic, suggesting that the high levels of this organism detected at the static pile site probably do not constitute a public health risk, however, the number of isolates actually tested for pathogenicity was very small relative to the quantity of *Yersinia* measured in some compost samples. These data do not definitively establish that virulent *Y. enterocolitica* were or were not present or that sludges do or do not serve as a reservoir of pathogenic *Yersinia* strains. Additional research will be necessary to fully address these questions and understand the ecology of *Yersinia* in affected sludge treatment systems.

Improving Composting Results

The data showed a higher occurrence of microorganisms in the static pile products overall. This should not be viewed as an indictment of static pile systems, per se. In fact, one of the static pile facilities sampled had very low densities of microorganisms. The reasons for these differences could not be determined from this study. The project author has seen a correlation between the amount of attention paid to the windrow composting process by the operators and the quality—microbiologically—of the final compost product. This suggests the need for good operator training and operation manuals.

While the research did not determine what caused the variations in pathogen *(Salmonella)* levels, it was subsequently found that more complete composting lowers overall microorganism densities and results in a compost product in which *Salmonella* regrowth is minimized. Monitoring for specific indicator microorganisms seemed to reflect this point. Therefore, monitoring or testing might be needed in addition to current guidelines to ensure that pathogenic bacterial levels are adequately reduced.

A voluntary coliform monitoring program to indicate the potential for *Salmonella* regrowth was instituted at the California windrow facility. In addition to meeting minimum recommended time and temperature requirements, compost there is not released to the marketing agent until total coliforms have been reduced to a median most probable number (MPN) of 10/gram dw. or less (based on recommendations published by Roger T. Haug in *Compost Engineering, Principles and Practice;* 1980). No more than 20 percent of the samples can have greater than 1,000 MPN/gram and no single sample can be greater than 10,000/gram. If the median number is above 10 per gram, composting of that windrow is extended for a week to 10 days, after which the material is retested. If a windrow does not meet the 10 MPN/gram median coliform level after two or three extensions, a test for *Salmonella* is conducted; to date, that has occurred very infrequently. The windrows are released if the results of the *Salmonella* test are negative.

In conclusion, the authors of this article are very encouraged that there were no significant problems with viruses and parasites—major groups of organisms that were thought could be a problem. Most of the major classes of pathogens that the Water Quality Laboratory looked for were absent. And although the study did find a potential problem with *Salmonella,* it appears to be a situation that is correctable.

William Yanko is Microbiology Supervisor at the San Jose Creek Water Quality Laboratory (part of the Los Angeles County Sanitation Districts) in Whittier, California. He is the author of the study reviewed in this article ("Occurrence of Pathogens In Distribution and Marketing Municipal Sludges.") John Walker is with the U.S. Environmental Protection Agency's Office of Municipal Pollution Control in Washington, D.C. Nora Goldstein is Executive Editor of BioCycle. *Walter Jakubowski is in the Toxicology & Microbiology Division of U.S. EPA's Health Effects Research Laboratory in Cincinnati, Ohio. Mr. Jakubowski was the project officer for the study reviewed in this article.*

18

REFINING RELATIVE RISK ASSESSMENT

Gene Logsdon

"WHEN zero risk assessment is employed or too rigid emphasis put on MEI (Most Exposed Individual) models, what can happen is what I call 'environmental gridlock.' In legislating for a totally risk free world, we end up with rules that prevent anything from happening," explains Ohio State University soil professor Terry Logan who focuses on sludge utilization.

Despite the complexity and unlimited variables of the biological world, scientists are guardedly optimistic about formulating models that will be helpful in assessing relative risk which will provide a framework wherein rules of reason can operate more effectively. Michigan State researcher Lee Jacobs, co-chairman (with Andrew Chang of the Univ. of Calif., Riverside) heads up one example of this kind of research on sludge utilization—a cooperative program known as W-170 after its project number. "W-170 involves agricultural research stations in the north-central and western regions, with Hatch Act funding through the Cooperative State Research Service," explains Jacobs. "The purpose of this funding is to encourage researchers to share common interests and to minimize duplication of effort." A 1988 meeting of the group brought together representatives from government, research, and municipal sewerage districts and was partially funded by USEPA.

W-170, along with its predecessor W-124, has for about seven years now supported experimental studies at various research sites throughout the country on the uptake of metals and organic compounds in plants. All cooperating stations use the same sludge—from Chicago—to reduce the number of possible variables. Originally, barley was the test crop grown, but in more recent years, the researchers switched to sudex. Its uptake of metals is greater, and it grows about the same everywhere, whereas with barley different varities were necessary for different regions.

"We have data already that challenges current thinking and will be of great help in making the first crude models for relative risk assessment," says Jacobs. "For example, uptake, say of cadmium, has been assumed to be linear, but our data indicates that after the initial period, uptake tends to level off—it is curvilinear rather than linear. Also under some conditions, applications of, say, 50 lbs. of a metal in one form may result in little or no uptake, where 20 lbs. in another form may result in a fairly significant uptake. When we get enough of this kind of information we can begin to formulate predictive models. They are going to be crude at first. But the more information you can put into the computer, the better the results, as we all know. This information can be more immediately valuable in helping us make *comparative* risk assessments, as between land application of sludge and incineration. Also, when we deal with organics, we're talking about thousands of compounds, unlike the six or eight metals we need to worry about. Here is where modelling will be of immense help even in the early crude stages, as a way to keep track of all the data."

It's going to be a slow process, emphasizes Jim Ryan of the EPA, who is concentrating on organics in his work. "There are thousands of organic compounds that need to be investigated. Picking the right ones to work with initially is important." Ryan is painstakingly measuring uptake of an organic compound present in soil and comparing that with uptake of the same compound present in sludge applied to the soil.

Ryan says there are two ways to choose which organic compounds should yield the most important information. One is by its relative concentration or occurrence. The other way is to look at the characteristics of the compound itself and study it independently. "Eventually though, whatever parameters of relative risk and safety we can arrive at, society must reach a decision on overall philosophy," he says. "What drives us bananas is that we can operate with a risk equation that makes sense to science only to realize that Congress is using a different kind of equation, one based more on perceived risk than real risk. For example, society has instituted self-imposed risks that cause a lot more harm and death than possible contamination from sludge, as for example raising the speed limit from 55 to 65. But the perception of risk from that is less than the perception of risk from sludge and that's what public officials and society react to. From a scientific view, there's not always much logic to it."

Despite nature's complexity, there is general agreement that, however monumental and ongoing the process will be, enough knowledge can be gathered to make reasonable relative risk assessments practical.

SECTION FIVE

MARKETS AND UTILIZATION

19

COMPOST FOR SALE

MORE than 215 municipal sludge composting facilities are either on line or scheduled to start operating in the future. In addition, a growing number of municipalities are making a commitment to compost solid waste. And there are literally hundreds of leaf and yard waste composting projects. Added together, there is a significant amount of organic material to be marketed to end users.

Most municipalities have had a great deal of success marketing quality sludge compost. A number of facilities, in fact, have more demand than supply. It is too early to analyze markets for solid waste composts. Many leaf composting programs give the product back to residents, use it for public works projects, or sell it in bulk to top soil companies, landscapers, and nurseries.

Because sewage sludge has dominated the composting field for the last decade, much of the current research and discussion with regards to marketing and utilization revolves around sludge compost. Yet the principles and experiences are applicable to most composts. This section starts with a discussion on marketing programs. Subsequent chapters address how to gain public acceptance and site facilities and plant and crop response to both sludge and solid waste composts.

While many municipalities continue to distribute the material themselves, others rely on private companies that are in business exclusively to market compost. This burgeoning "industry" is a direct reflection of the output of new facilities generating material.

Some cities—such as Hampton Roads, Virginia and Durham, New Hampshire—find it effective to handle compost marketing and utilization themselves. In these cases, they look at the "in city" potential use by street and parks departments. An examination of how much a municipality pays for fertilizer, topsoil and peat indicates the potential demand for the compost.

Other cities, like Scranton, Pennsylvania, take a dual approach—contracting with an independent firm to market some of the product, while allocating a percentage to utilization by public agencies.

"A city needs to evaluate the versatility of its equipment, specifically the suitability of the equipment for applying compost to city parks and so forth," says Ron Albrecht of Ron Albrecht Associates in Annapolis, Maryland.

When Maryland Environmental Service opened its composting facilities in 1980, it opted for establishing its own marketing program. The revenue from the sale of the compost offset the administrative and promotional costs. The City of Philadelphia, on the other hand, felt it didn't have the inhouse marketing expertise or a suitable infrastructure to administer a program. So rather than start from scratch in developing a program on its own, the Water Department decided to enlist the services of a professional marketing company, similar to the relationship of long standing between the Los Angeles County Sewage District and the Kellogg Corporation.

Denver, which opened a full-scale composting facility

in 1986, chose to do its own marketing. The city has a dual sludge utilization program where 60 percent of the annual sludge volume is land applied and 40 percent is composted. "We get to keep more control over the operation," explains Robert Tardy of Metro Denver, "especially because we can coordinate it with our system here. We didn't want someone out there selling compost we didn't even have." To date, Denver has been selling all of its compost to a wholesaler who reprocesses it into a potting soil.

Cities with their own programs enter into the sensitive area of being in business in the public sector. "Municipal employees who are responsible for marketing compost to chain stores and nurseries can be at a disadvantage with salespeople who work for private firms when it comes to flexibility in dealing with potential customers," says Albrecht. "That's one of the realities of getting orders."

Other constraints concern pay scales and authority, he adds. "A city is generally very limited in what salary it can offer to experienced marketing personnel. Also a city or county may not have the legislative authority to get into the business of marketing compost products. Finally, if and when a city succeeds in a marketing effort, it may create problems in competing with established fertilizer or soil conditioner firms."

Ultimately, there is no single, set solution for all municipalities. As more and more choose the composting option, they will have to evaluate the scope of their own potential uses relative to the amount of compost that will be produced, and how well its own resources—personnel as well as physical equipment—can support a marketing effort. And with the existence of a compost marketing industry, municipalities have a choice of which method best suits their needs.

Marketing Case Studies

The Hampton Roads Sanitation District in Virginia Beach, Virginia began operation of its 12 dry ton/day Peninsula Composting Facility in 1981. Sludge from two of the District's wastewater treatment plants is composted using the aerated static pile method. The District chose to market the compost itself, and initially distributed the material at no charge, to introduce the product to potential users. A pricing structure, which has been modified several times, was subsequently established for bulk sales.

By 1986, sales of the compost—with annual production of 12,000 cubic yards—were as follows: 42.5 percent landscapers; 36.2 percent public; 13.3 percent nurseries; 7 percent municipal; and 1 percent retail. Most of the landscapers—about 90 percent—use the compost to establish turf, either to install new lawns or do reclamation work, says Rhonda Oberst, marketing specialist for the District.

There were about 100 days in 1987 where the District had to stop sales because there was no compost available. Prices in 1988 for the product (bulk sales) were: $12 for under 10 cubic yards, $8 for 10 to 499 cubic yards, and $6 for purchases of 500 cubic yards and over.

To better serve the public market, as well as nurseries, the District began a bagging operation in 1986. Market research had shown that homeowners were interested in buying compost, but in quantities less than one cubic yard. In addition, retail outlets had begun expressing interest in carrying bagged compost.

The District recognized the need to design a bag that would attract users and distinguish the product from its competitors. Thus capital for the bagging operation was invested in bag design and not equipment. The District also decided that it would sell bags to wholesale distributors, rather than directly to consumers. The compost currently sells for $1.35/bag or $67.50/pallet (with a minimum purchase of two pallets). The average retail price is $3.50/bag. The product is sold in about 30 different locations. Retailers are provided with literature on usage.

Stores that buy the two pallet minimum receive a free ad in a flyer that the District distributes via its sewage bills. "The ad lists the name, address, and phone number of the outlet," says Oberst. "The District includes the ad in the bills twice a year—in the spring and fall—which are the two big seasons for compost use. The mailer is one of the best ways to let people know compost is available and where they can buy it."

Over 10,000 bags of compost were sold in 1987. The District recently invested close to $100,000 in a prefabricated building and bagging-related equipment, e.g. a conveyor, bag sealer, forklift, and pallet wrapper. "This is where we will be concentrating our marketing effort for the next few years," says Oberst. "I feel we can reach out to a whole new user group with them. And there is a much greater profit margin on the bagged versus the bulk compost—$36/cu. yd. vs. $12/cu. yd."

In addition to targeting retail stores, the District believes that nurseries may be more receptive to that kind of packaging. "Bags would be easier for nurseries to use," adds Oberst, "because they could be stored more conveniently."

The District has the capability to meet the needs of both the landscaping and nursery industries by varying the screen size. A 6 mil screen can be used to produce a coarse product for potting mixes. Landscapers and golf courses need a finer textured product for turf establishment, so the screen size can be switched to a 4 mil size to accommodate these groups. Currently, the composting facility is only producing the finer compost, since landscapers are the primary market.

The District currently spends about $15,000 a year (not including salaries) for its compost marketing program. In 1987, operating costs at the composting facility were $177/dry ton. The net cost, after revenues from compost sales were factored in, was $143/dry ton.

The Hampton Roads Sanitation District has a successful compost marketing program. The number one key to success, says Oberst, is having a positive attitude—a recycle/utilization versus a disposal mentality. "If you're out there with the general public, don't use words like 'dispose of' or 'get rid of' the waste."

The second tip is to research all markets in your area, instead of sticking only with familiar ones. The third is to find a university or local agricultural station to work with your compost product. This builds third-party credibility for the compost.

Finally, price your product with a value, but don't start too high and be forced to mark it down. "Choose a comfortable price and then increase it with demand," concludes Oberst.

Private Sector Marketing

Kellogg Supply in Carson, California markets compost from the Los Angeles County wastewater treatment plant's windrow operation. The company, which has been in business for more than 60 years, has 12 blends of fertilizer and seven different soil amendments that all incorporate the compost. The products are sold primarily in nurseries and chain stores in southern California, although Kellogg is expanding into northern California, Nevada, Arizona and Hawaii.

Across the country—in Nashville, Tennessee—a far newer company is developing, relying on bulk wholesale marketing to move product from the city's treatment plant. "Only 20 percent of the material we sell is straight compost," says Odell Binkley of Soil Products, Inc. "Eighty percent is blended and mixed." The Nashville facility uses aerated static piles and expects to process 40 dry tons per day.

While bulk sales go mostly to nurseries and landscapers in the Nashville area, Binkley says some shipments go to Louisville, Atlanta and Huntsville, Alabama. Recently the Soil Conservation Service accepted compost in the USDA cost-share program. As soon as the specifications for compost are written at the University of Tennessee, farmers may make up more of the market.

GroCo, Inc., in Seattle, Washington, has a slightly different arrangement. The company, which got its start composting sawdust and steer manure, has a contract with the City of Seattle to compost and market about 10 percent of its sludge. Landscapers and the general public are the primary users of the product. GroCo also has a pilot-scale yard waste composting project. There are a growing number of companies that compost leaves and yard waste, charging a tipping fee to municipalities and counties. This trend is expected to continue, especially as more states ban the disposal of these materials in landfills.

One of the most aggressive marketing companies is Compost Management, Inc. in Doylestown, Pennsylvania. Its Earthlife brand is used to market sewage products from treatment plants and MSW compost from solid waste facilities, not just in Pennsylvania, but as far away as Florida. Roger Tuttle, president of Compost Management, believes that all on spec, nonphytotoxic compost can be sold—although the task is time consuming and complex. "We forecast that the price paid to the compost plants will be between \$1 to \$4 per ton," he says. "The price will not be higher because of the amount of effort and steps required to market the compost—from conducting greenhouse research and phytotoxic testing to convincing growers how to use the compost properly, advertising and public relations, attending trade shows and holding demonstration days and developing the right spreading equipment—the price cannot go any higher. If the price were raised, the compost would not be competitive in the marketplace and volume would be reduced."

The company has 22 major end uses for compost, identified by the largest to the smallest number of customers in each category. "You must know the location of these customers in relationship to each plant because the most sensitive factor in compost marketing is transportation. You will have a transportation dislocation problem more quickly than you will have a problem with product pricing. Regarding order size, we will not get involved with anything less than a 40 or 50 cubic yard truckload. We market through a national franchised network of dealers who have been taught our concept of marketing."

Tuttle is succinct when it comes to the question of necessary product quality controls: pathogen, heavy metal and PCB controls are the most important, he says, and if there is no pretreatment program, "don't even think about composting." He explains how he has come to find out the necessity for complementary expertise when marketing compost.

Product Moisture—"You need to know whether commercial equipment can utilize a compost with a 55 percent moisture level and be able to blend it and be aware of what problems will occur if the percentage is higher or lower."

Organic Matter Content—"Originally we thought compost was a 100 percent organic material, and obviously it is not. It's half organic, and half mineral matter. The mineral matter is the size of sand, silt and clay. The structural and textural analysis of compost relates directly to specifications for many landscaping jobs."

Grower Acceptance—"Probably the hardest end use to sell is the growers. Compost is the least cost raw material to the grower even though he is adding as much as 30 percent compost to his mix. The seed costs more, as does the plant stock and fertilizer. The grower is very cautious to change his formula. Therefore, the correct marketing approach is imperative."

Product Variations—"We continually try to understand compost quality. How is it different from facility to facility? Lime dewatered, polymer dewatered, hardwood chips, softwood or sawdust? What kind of agronomic difference is created in final product when the source material changes from solid waste to sludge. What are the phytotoxic leachate characteristics of each compost. These are just several of the many factors that are needed to establish a viable marketing program."

Nursery and Landscaping Markets

Frank Gouin, a professor of ornamental horticulture at the University of Maryland, has been working with sludge composts since the early 1970s. He has found, for example, that compost can be used successfully in container mixes in the greenhouse industry, reducing the need for fertilizer by at least 50 percent. The rate of plant growth is accelerated. The amount of peat in a mix can also be reduced. In the landscape industry, compost can be substituted 100 percent for peat or pine bark and, says Gouin, there is no need to add fertilizer.

"The market is there to absorb compost," says Gouin. "The horticulture nursery and landscaping industries utilize more organic material in a raw state—in container mixes and as a soil amendment—than any agricultural industry. They will be the biggest users of compost."

While penetrating that nursery-landscaping market has been slow, attitudes are beginning to change. Gouin cites several factors: 1) significant research that shows the material can be used very effectively; 2) financial savings by using compost; and 3) a belief that they are doing something environmentally beneficial.

When selling to this market, however, quality control—consistent product quality—is absolutely essential. "One thing people have to understand is that once they get the product on the market it has to be the same week after week, month after month, year after year," says Gouin. "You are selling to growers who have a minimum of $20,000 invested in one acre and a maximum, depending on the species being grown, of $45,000. It is one of the most intensive industries. So quality control is very important. You have to guarantee your product. The industry is taking a big gamble using compost, given that investment."

Being certain of consistent quality is one of the essential steps to marketing successfully. Some of the others, points out Gouin, include:

• **Before selling any compost that's produced, test it.** Different composts need certain testing to establish the background information, e.g., if it is lime dewatered, it can't be used on certain kinds of plants.

• **Emphasize the fact that you are marketing a clean sludge.** "People have to understand that sewage sludge is cleaned up," says Gouin. "Composting is only being done on the cleanest of sludges."

• **Provide users with very specific instructions.** "If compost is to be used effectively, certain steps must be followed," says Gouin. "Compost is different than materials they have been working with, and they have been doing things a certain way for years. Growers need to know, for example, the concentration of the mix, the crops the compost has been tested on, and the fertilizer needs. Therefore, you must come in with a strong definite plan and emphasize that there is no other way to use the material."

• **Get potential users to try the compost—but make them invest some money from the start.** "If you give them a truckload of compost to try, charge them for the transportation costs," says Gouin. "Then the next time, if they like it, they'll pay for the load. If you give it to them absolutely free of charge, they haven't invested anything initially and then the material isn't worth anything to them."

Eventually, he adds, compost can be priced competitively with the other organic materials on the market. Initially, however, it may be necessary to price it lower in order to get people to try it.

• **Focus on the innovators in the industry.** "There are three categories of users in the industry—the innovators, the people in between who might change once they see someone else change their practices and make money, and the people who won't change no matter what happens," he says. "You must distinguish between those three categories and then concentrate your marketing efforts on the innovators. They are usually willing to listen, and you can show them compost is a resource, and guide them through changing their methods of growing."

While there may be a market for any kind of high quality compost, certain composting practices expand the applicability of the material. Gouin discourages the lime dewatering of sludge if the compost is to be marketed to the horticultural industry. In terms of particle size, no material over a 1/2-inch diameter should be allowed in the compost after screening. And most importantly, says Gouin, the compost should be aged adequately.

"Compost improves to about four months," he says, "and after that, it's at its best. I advise growers to buy compost a year ahead of time and cover and store it. The older the compost gets—the longer the curing time—the more consistent the results I get in terms of plant response."

Compost produced via the aerated static pile method usually has a long curing time. Some of the enclosed reactor systems, however, have a shorter curing/retention time. "In those cases, when the compost is being sold, the grower should be told not to use the material for six months," says Gouin. "This relates back to the issue of quality control because not only does the material have to be stockpiled, but it must be stockpiled correctly.

"I continually emphasize curing and careful stockpiling. But the big problem is that many people make the piles too high and then the pile goes anaerobic, forming alcohol inside. That material kills plants. When storing compost, build the piles in a cone shape not more than 10 feet high and the bottom not more than 20 feet across. If you modify that in any way, you need to aerate the piles."

Gouin continues to do some research on the use of compost in the horticultural industry. Research funds, however, are limited. "The hard core work has been done on greenhouse and nursery crops," he says. "Now it's a matter of changing the attitude of potential users. And that's where the marketing comes in."

20

HOW TO SUCCEED IN COMPOST MARKETING

Ronald Albrecht

INVARIABLY, whenever a new composting project is considered, this question comes up: "What are you going to do with the compost after you make it?" This question is followed by "first hand" accounts of unsold compost stored at some facility.

This skepticism is understandable if you are aware of the compost marketing problems of the 1960s. Today, however, we find that compost is successfully marketed throughout the United States. At many locations, the product is sold out and the market demand remains unsatisified.

But by the end of this decade, almost 200 sludge composting facilities should be in operation. When all of them are operating, there will be increased competition for markets. This is especially true in the Northeast and mid-Atlantic regions where the most rapid rise is predicted. Marketing areas will overlap and new strategies will be required to sell the greater production.

Marketing sludge compost differs from marketing other humus products because of the original source of the material and constraints placed on product use. Since the reuse of any wastes, especially those containing human wastes, carries a stigma in the minds of many people, there is always a psychological barrier. This barrier cannot be avoided but can be overcome using consumer education and demonstration of product value.

Other concerns such as regulatory interest in both public health and environmental risks of product use must be acknowledged and accommodated. Regulations on product use must be adhered to in all phases of marketing.

In the case of commercial users, the results of using the product directly affect the growers' profits. In some areas there are political concerns. Many times the political concerns are unspoken and manifested as restrictions on use of compost or long delays in granting permits for distribution.

Compost marketing is now entering a maturing phase and has developed some key essentials common to all good programs. These are not the only ingredients in a program but are present in successful ones.

- The product should have consistent quality. This is very important in all sectors of use but is most significant in commercial operations where predictable results and repeated sales are related.
- The supply of the product must be reliable. Commitments to buyers must be fulfilled. Inability to meet market commitments causes bad customer relations and destroys the credibility of the marketing program.
- Production of a variety of products increases the success rate of marketing. Specialty products can be made by adding other sources of nutrients, mixing with other materials to make potting soils or artificial topsoil, or physically changing the product to broaden the marketability of the product. Marketing a variety of products of different values can also produce better revenues and increases the probability that all product will be sold.

• The marketing institution must stand behind the product, maintain credibility and integrity. This means that in the event of customer dissatisfaction with the product, the results of product use or the marketing service, you give the customer the benefit of doubt, accept responsibility, immediately follow up on complaints, and promptly take corrective action. This is an important policy to follow in developing a good marketing image for a new product.

• There are no "pat" or set solutions for marketing compost which are applicable to all situations. Differences in products as well as differences in soils, agronomic practices, social patterns, and climate create a new set of variables for each project. The marketing plan should represent the best fit to meet the site-specific situations.

• Marketing is a continuing effort, it is not just "taking orders." Aggressive marketing is necessary to assure that all product is utilized and that revenues are optimized.

Inventory the Potential

The first step in developing a marketing program is to inventory existing and potential markets, quantify all of the markets, and develop target markets. This involves knowledge of the product, product use, constraints on use and estimating the value of the product to the user. The assembled information can then be analyzed to develop a marketing strategy, management structure and management plan.

It is important to keep in mind that a marketing plan is not an inflexible guide and should be adjusted to changes in the markets. Marketing plans should be reviewed on a frequent basis in order to anticipate and adjust to market changes. Without flexibility and frequent review, a marketing plan can become a road map to failure. This can easily happen when the review of marketing performance is *limited to a review of the revenues.*

Revenues are, in a sense, the results of past efforts and may not accurately predict future performance. This is especially true in new industries where changes in regulations and reactive changes by competing products are expected. There is now a trend to peer review of marketing efforts in order to insure that the marketing results meet the industry norm, the marketing plan contains the best program, and the management structure is the most effective to achieve the project goals.

In summary, compost is and can be marketed throughout the country. The primary goals of reliably disposing of the product are being met. The need to optimize revenues and the increased competition are becoming important factors in the management of marketing programs. Aggressive marketing and the ability to market more than one product offer the most promise as ways to enhance revenues and maintain a share of the market without sacrificing the reliability of disposal. The compost marketing industry is maturing and successfully adjusting to meet new challenges.

21

COMPOST VALUE TO ORNAMENTAL PLANTS

James B. Shanks and Francis R. Gouin

THE authors have used processed sewage sludge products since 1973 in attempting to find satisfactory means of incorporating them into media for the production of container grown floricultural and ornamental plants, and in the production of nursery liners[1,2,3,4]. The potential danger of the accumulation of heavy metals from continued use of such materials need not be a concern in the production of ornamental plants.

Dewatered raw sewage sludge (20 percent solids) was mixed with twice the volume of mixed hard-soft wood chips and composted in piles for 30 days under controlled aeration. Temperatures of 80°C or higher were attained during composting and most pathogenic organisms destroyed[12]. Following an additional 30 day curing the wood chips were removed using a 1.25 cm rotary screen. Analysis of this raw sludge compost (hereafter referred to as compost) has been reported as: pH 6.8; water 35%; carbon 23%; total N 1.6%; NH_4^+—N 235 ppm; P 1%; K 0.2%; Ca 1.4%; Zn 770 ppm; Cu 300 ppm; Cd 8 ppm; Ni 55 ppm; Pb 290 ppm.[12]

Using one-third by volume compost, nine root media were prepared with equal parts of two other components commonly found in root media:

1) silt loam soil.
2) horticultural grade peat moss.
3) pine bark milled to pass through a 1.25 cm screen.
4) perlite — a heat expanded siliceous volcanic rock, horticultural grade.
5) vermiculite — a heat expanded, micalike, silicate mineral, horticultural grade.

Media of equal parts by volume of peat, bark, and perlite and of peat, bark, and vermiculite each with ground limestone, but with and without the fritted trace elements (FTE'), were used for comparative purposes. A commercially available medium of peat, perlite, vermiculite, and sand *(Metro Mix 400')* was included for comparison. Plants used in the experiment were lilies, poinsettias, and chrysanthemums.

Results and Discussion: All plants resumed good growth after planting in all media indicating that any potential root injury from the salt content of the compost amended media (Table 1) was avoided by leaching at the first watering. The damage from salinity should have also been lessened by media containing soil, and to a lesser extent bark, because the initial conductivity of these soil extracts was less than with peat, presumably due to an absorption of ions by these materials. It should be noted that the conductivity of many of the compost containing media was in the same range as the *Metro Mix 200* medium which contained an undisclosed amount of nutritional elements.

The acidity of compost containing media was quite stable, being at or close to the neutral range both before and after cropping. The apparent uniformity at the end of the period of growth may have been due to the larger number of replicates. There was no indication that the continued use of fertilizer during the cropping period increased acidity, indicating that the relatively neutral nature of the compost was highly buffered. The media prepared with ground limestone actually tended to become more neutral (higher pH) during the period of plant growth.

The air-filled pore space (APS) at the end of the growing period was quite variable, being generally greatest in compost containing media made up with peat, perlite, or bark, in that order. Compost media with bark and *200,* in peat, bark, and vermiculite, and in media of compost with bark and vermiculite, peat and vermiculite, peat and perlite, and with soil and vermiculite. Plants in peat, bark, and perlite were delayed in blooming. There were no differences in plant height. Chrysanthemums thus grew satisfactorily in a wide range of media but did particularly well in media containing vermiculite.

Lily: Stem length in lily indicates good growth but the most important factors in commercial production are the number of flowers and the number of days required for flowering. The tallest plants were produced in media of compost with soil and bark, with perlite and vermiculite, and in peat, bark and perlite. The only plants that were significantly shorter were those grown in compost, soil and vermiculite and in compost, peat, and vermiculite. The greatest number of flowers was on plants grown in the medium of compost, peat, and perlite although the number of flowers was not significantly different on plants grown in media of compost with soil and perlite, peat and vermiculite, bark and perlite, bark and vermiculite or in the media of peat, bark, and perlite with FTE, peat, bark, and vermiculite, or in *Metro Mix 200.* Earliest flowering was on plants grown in media perlite, peat and perlite, and with peat and bark had the highest APS, being just slightly less than the APS of peat, bark, and perlite. All compost containing media without soil had as much or more APS than the commercially available *Metro Mix 200.*

Poinsettia: The mean fresh weight of all cultivars of poinsettia was greatest when plants were grown in peat, bark, and vermiculite amended with FTE, in *Metro Mix 200,* in peat, bark, and perlite, and in compost, bark, and perlite media. The number of flowering branches, a measure of ornamental value, was greatest in media of compost combined with soil and bark, peat and vermiculite, and with bark and perlite. Plant height was not affected. Thus, the best overall growth of poinsettia was in the medium of compost, bark, and perlite.

Chrysanthemum: Chrysanthemum had the greatest fresh weight grown in a medium of peat, bark, and vermiculite amended with FTE, in *Metro Mix 200,* and in compost media combined with soil and vermiculite, peat and perlite, peat and vermiculite, bark and vermiculite, or with perlite and vermiculite. The most flowers were produced on chrysanthemum plants grown in compost, perlite, and vermiculite, in *Metro Mix* of compost with peat and vermiculite, peat and bark, perlite and vermiculite, and in peat, bark and perlite with FTE.

General: Our data has indicated that compost was satisfactory for use as one-third by volume of the root media in combination with several other materials commonly used in a soil based or a soil-less medium. The combination of compost with bark and perlite was particularly satisfactory for poinsettia. Chrysanthemums had a wider latitude for satisfactory combinations of ingredients but the inclusion of vermiculite was desirable whether the third ingredient was soil, peat, perlite, or bark. Lily grew best in media containing perlite or bark. There was no indication that liming or additional micro-nutrients would be beneficial to any of the three species growing in a medium of one-third compost.

References

1. Gouin, F.R. 1977. Conifer tree seedling response to nursery soil amended with composted sewage sludge. *HortScience. 12*(4):341-343.

2. __________. 1977. Using sludge to improve the soil. *Amer. Nurseryman CXLVI*(5):15,110-115.

3. __________ and J.M. Walker 1977. Deciduous tree seedling response to nursery soil amended with composted sewage sludge. *HortScience. 12*(1):45-48.

4. Shanks, J.B. 1974. The use of several industrial municipal organic by-products as components and plant growth media. *The Maryland Florist 192*(July):1-7.

5. Willson, G.B., J.F. Parr, E. Epstein, P.B. Marsh, R.L. Chaney, D. Colacicco, W.D. Burge, L.J. Sikora, C.F. Tester, and S. Hornick 1980. Manual for composting sludge by the Beltsville aerated-pile method. Municipal Environmental Research Laboratory, Office of Research and Development. U.S. Environmental Protection Agency. Cincinnati, Ohio, U.S.A.

22

SUPPRESSING PLANT PATHOGENS WITH COMPOSTS

ACCORDING to Harry Hoitink, professor of Plant Pathology at the Ohio Agricultural Research and Development Center, a significant body of knowledge has been published on biological control of soilborne plant diseases with composts. This has had an impact on procedures used by the ornamentals industry, particularly for production of containerized plants.

Traditional plant growth media consisted of peat, sand and vermiculite or Perlite. These media were treated with aerated steam and drenched routinely with fungicides because they are conducive to disease development.

In the nursery industry in particular, but also for production of floral crops, composts prepared from tree barks and other organic wastes have partially replaced peat in growth media. Some of these media suppress several important soilborne plant diseases. This has eliminated the need for steaming and reduced or negated fungicide use.

In a 1988 interview with *BioCycle,* Dr. Hoitink was asked if finished compost from sludge suppresses soilborne diseases as well as hardwood bark compost, which was the subject of his earlier research. His response was as follows:

"The heat of composting not only kills fecal pathogens and plant disease pathogens and even tomato seeds for example, if the process is performed properly, but beneficial microorganisms antagonistic to the plant pathogens are also destroyed. To renew the compost's suppressive "power" over plant pathogens, it needs to be reinoculated with the beneficials. From our exhaustive experimentation, both in the laboratory and in industry, we've learned that if we allow natural recolonization to take place, it proceeds most of the time in a haphazard manner in which one plant disease may be suppressed and not another, or maybe none. This is especially true for the new in-vessel composting systems. So what we had to do was identify beneficial and detrimental microorganisms, how they act under varying conditions, and how and when to reintroduce the beneficial ones into the compost."

Dr. Hoitink's research has been greatly assisted by a new instrument — the 5898A microbial identification system from Hewlett-Packard — which very quickly identifies known microorganisms in samples fed into it. Explains Dr. Hoitink:

"What we have been able to do is isolate certain microorganisms or groups of microorganisms that if reintroduced into compost before natural populations of plant pathogens get established, make the compost very effective in suppressing Rhizoctonia and some other pathogens that cause damping off in seedlings. Effective enough, in fact, that we have been granted a patent for the production of disease suppressive composts and the microorganism culture for use in them. Our research is ongoing. For example, with the flats of radish plants in the lab, we isolate microorganisms, then inoculate them into the potting medium, or dip the seed in the inoculum, and see what happens to the radishes. Differences in suppressive or conducive ability can be quite dramatic."

Concludes Dr. Hoitink about the significance of utilizing compost as a protective material for plant growth:

"When we can solve a waste problem by turning it into a real asset for horticulture or agriculture, and at the same time avoid costs involved in heat sterilization or possibly hazardous chemicals, that really is progress."

SECTION SIX

CO-COMPOSTING OF SOLID WASTES

23

AN INTEGRATED RESOURCE RECOVERY SYSTEM

Luis F. Diaz, George M. Savage, Clarence G. Golueke

A combination of waste management subsystems can be used to achieve highest reduction at the lowest economic and environmental cost.

Recyclable materials are ferrous metal, glass, aluminum, plastics and paper fiber. Alternatively, the paper fiber and/or plastics can also be processed into a refuse-derived fuel. In the system, these various fractions are isolated along with a highly organic fraction suitable for composting alone or with sewage sludge.

To compensate for periods of unfavorable market conditions, system diversity is an absolute necessity. A cardinal rule must be that a waste management system cannot shut down when the market for a particular product has been temporarily curtailed. The integrated system provides for direct recovery of 60 percent of the incoming solid waste stream in the form of recyclable secondary materials, along with the capability of increasing recovery to 80 percent using additional processing subsystems.

The need for reliable information on the quantity and composition of the waste stream is well documented.[1,2,3] In the past, the lack of attention to detail paid to the types and characteristics of solid waste have, in fact, substantially limited opportunities for material recycling. Equally important, integrated waste management systems are doomed to failure if designed with quantity and composition data that are insufficient, inappropriate, or both.

For the purpose of this presentation, the average of actual waste characterization studies conducted throughout the U.S. has been used.[4] It would be entirely inappropriate to attempt to use this composition for any actual locality. In order to simplify this analysis, the various components have been combined into five major categories. The removal of paper, glass, metals, and "other organic" materials would leave a residue consisting of basically inorganic matter of about 20 percent of the total waste stream.

The system described herein is of modular construction, which allows for several different or combined processing subsystems. The basic module is the dry processing plant, which provides the intermediate products for further end-product enhancement. In the dry plant, ferrous metals, light and heavy organic fractions, and a clean light fraction (a mixture of predominantly paper and plastic) are recovered. The ferrous metal fraction is of sufficient purity to be sold as ferrous scrap.

Glass and non-ferrous metals can be easily (and more efficiently) removed from the waste stream by separating them at the source and collecting them at the curb or transporting them to a recycling center. The screened light fraction provided by the basic dry plant can be used, without further processing, as refuse-derived fuel, or as a raw material for a wastepaper pulp mill. The light fraction can also be further processed to provide a highly homogeneous refuse-derived fuel if required by the user. The latter two alternatives represent additional modular processing subsystems.

Lastly, the organic fractions can be used for composting. The nutrient content of compost from refuse can be greatly enhanced by mixing the refuse with sludge or agricultural wastes. Sewage sludge can be added in amounts equivalent to the daily per capita production.

Process Modules

The basic dry plant consists of primary size reduction, air classification, and screening. The heavy and light fractions from the air classifier are processed separately in the "heavies" and "lights" recovery systems. The heavies undergo ferrous removal.

There are various processes that can utilize the screened light fraction recovered in the basic dry plant. First of all, being high in paper content, the screened lights can be treated as a furnish for a fiber recovery system. This system is similar to a conventional wastepaper pulp mill which embodies several fundamental processing innovations to remove the contaminants. It also includes a wastepaper treatment system where all wastewater is treated and recycled back into the fiber recovery system.

The screened light fraction can also form the raw material for the preparation of a high-quality refuse-derived fuel. The flexibility of the system enables the production of RDF in the form of a low density "fluff" or a densified pellet depending upon the particular requirements of the combustion unit. The RDF is compatible with many combustion technologies now fired by woodchips, coal, fuel oil, or natural gas. The heating value of the RDF is approximately 6600 BTU/lb, or about 40 percent of No. 6 fuel oil. The RDF has a sulfur content in the range of 0.1 to 0.3 percent and, when burned, has an ash content in the range of 10 to 15 percent. Hence, the RDF is an excellent non-sulfur, low-ash substitute for many conventional fuels.

In the system, the light and heavy organic fractions provide the feedstock for the composting system. After removal of the majority of the inorganic materials, these two fractions are combined and mixed with sewage sludge and are composted.

Environmental Impact

The integrated system seeks to optimize resource recovery from solid waste. The degree of material recovery from the waste stream depends upon many conflicting criteria, of which economics and technology are only two. At this point in time, technically 80 percent of all urban solid waste can be recycled or converted to energy using the integrated system approach described herein. The remaining 20 percent (composed of low value materials such as leather goods, rubber and rags, and a small percentage of inorganic materials) ultimately would have to be disposed of in a landfill. As resource recovery technology advances and markets become available, the size of this residual fraction may be further reduced.

In those modules which require water for processing, namely the fiber recovery system, the wastewater either is treated and recycled, or is treated and released. The quality of any process water released from the plant would be within the standards set by the governing regulations.

Precautions must be taken to control particulate emissions generated during the dry processing operations within the resource recovery plant. Consequently, suitable dust collection and suppression equipment, such as inertial separators and dust hoods, are used where needed in conjunction with a bag house.

Since all the recovery processing equipment is located inside a building, there would be no noise pollution problems. Although the noise level within the building may reach 95 decibels, the noise level outside the building remains below 40 decibels. Plant workers would be required to wear ear protection whenever in the vicinity of noisy equipment.

Co-Composting

The concept of co-composting is as old as that of composting municipal refuse. The difference between the "old" (pre-1975) and the "modern" (post-1975) concepts is one of emphasis. Originally, the emphasis was on the refuse fraction, and the sludge was regarded principally as a source of nitrogen and phosphorus and, to a lesser extent, of potassium for enriching the refuse. A seriously limiting factor prior to 1975 was the high moisture content of the sludge. At that time, the technology of sludge dewatering was in its infancy and, hence, sludge underwent very little dewatering (concentration) before being added to the refuse.

Co-composting systems in use in the pre-1975 years were earlier versions of those presently available. Thus, both static (forced aeration) and turned windrow systems were used, as were various proprietary reactors.

The requirement for bulking can be bypassed by co-disposal of the sludge with the organic fraction of refuse. The mixing of the two waste streams would result in a mixture that would have an average moisture content of between 30 and 40 percent. The resulting refuse/sludge mixture has the physical and chemical conditions appropriate for stabilization through composting. Indeed, the technical feasibility of composting different mixtures of refuse and sludge was demonstrated in the U.S. and in other countries many years ago. Furthermore, the economics of disposal would benefit from the fact that the simultaneous processing and treatment of refuse and sludge requires only one facility.

Even though it could be assumed that composted wastes, in addition to being an excellent soil conditioner, have relatively low concentrations of toxic substances and are free of pathogenic organisms, a large municipal compost operation would, nevertheless, have to surmount the major obstacle of developing a market large enough to accommodate the output. Co-composting refuse and sludge results in the production of much larger quantities of compost than would be produced by solely composting sludge. Indeed, a plant designed to compost the wastes from a municipality of one million people could produce about 600 tons of finished compost each day.

Except for a provision for the addition of sewage sludge, the steps in co-composting, regardless of compost method used, are the same as for composting any other organic residue. In this particular case, the sequence is as follows: collect and process the yard waste, segregate the organic fraction of refuse, mix the materials with sludge cake (or vice versa), compost the mixture, prepare the product for storage, use, or sale.

With respect to sludge, the principal requirement within terms of the compost process itself (i.e., exclusive of the use of the product) is a thorough knowledge of the characteristics of the sludge (e.g., organic content, moisture content, C/N ratio, P content). These characteristics have a decisive bearing on the suitability of methods and equipment chosen to handle and process the sludge.

Even more so than with the sludge, and for the same reasons, a thorough understanding must be had of the physical (particle size, moisture content, etc.) and chemical characteristics (e.g., N, P, K, C) of the yard waste and of the organic fraction of the solid waste stream.

Mixing Refuse, Yard Waste, and Sludge Cake

Mixing involves two important considerations, namely: 1) the proper ratio of sludge to other wastes; and 2) the act of mixing the wastes to form a homogeneous mixture. Many factors enter into a determination of a proper or suitable ratio. Among them might be whether the disposal of refuse supersedes that of sewage sludge, or vice versa. If refuse disposal supersedes sludge disposal, then the ratio of refuse to sludge would be as high as consonant with successful composting. If sludge disposal is the cause of greatest concern, then obviously the reverse would be true. However, if the disposal of both is of equal concern, then the ratio should be such that the relative generation rates of both wastes would be matched. Fortunately, all three of the general ratios can be used with equal success by suitably adjusting the moisture contents of the constituents, and taking care that the C/N ratio of the mixture is at a permissible level.

A thorough mixing of refuse with sludge cake (or vice versa) is essential to the success of a co-compost operation. Unfortunately, it also is one of the more difficult tasks to accomplish. Ideally, the mixture should be completely homogeneous, and yet be porous. Moreover, this porosity should be maintained throughout the duration of the composting process. Although the addition of yard waste contributes to maintaining porosity, a mixture of refuse and sludge alone has an innate tendency to settle, i.e., compact, whether it be in a windrow or in an enclosed reactor. The tendency is not surprising in view of the fact that refuse in the U.S. is largely paper in composition, and sludge cake is on the order of 70 to 80 percent water.

The viscous nature of the sludge cake (comparable to wet clay) renders homogenizing it with refuse a particularly difficult, and onerous undertaking. A major difficulty is the tendency of the two wastes to agglomerate into large clumps and globose masses.

Although an obvious solution would be to use dried sludge, the mixing problem can be met with a machine designed along the lines of the one used in studies by the authors.[9] The tendency to settle could be lessened by repeated turning, but not if a front-end loader is used. The use of forced aeration might also be a good approach, provided, of course, the particle size of the refuse is such as not to promote channeling of the air.

Inasmuch as equipment and processing involved in the composting and product preparation steps are the same as those used in the composting of refuse alone, and an abundance of literature is readily available on the subject, no further consideration is accorded them in this paper.[6,7,8]

If the entire operation has been properly conducted, the end product will have all of the well known qualities attributed to a good compost. However, this optimistic blanket statement must be qualified. Unless extraordinary measures are taken, the product will be blemished by the presence of glass shards and plastic particles from the refuse contribution and by heavy metals and toxic organics from the sludge, albeit somewhat "diluted" by the refuse. Fortunately, with the increasing sophistication of equipment and technology, a way can be found for avoiding or removing these blemishes.[10]

The authors are on the staff of Cal Recovery Systems, Inc. of Richmond, California.

References

1. Savage, G.M., and L.F. Diaz, "Key Issues Concerning Waste Processing Design," Proceedings of the 1986 ASME National Waste Processing Conference, Denver, Colorado, June 1986.

2. Diaz, L.F., G.M. Savage, and C. G. Golueke, "Production of Refuse-Derived Fuel from Municipal Solid Waste," presented at the 79th Air Pollution Control Association Annual Meeting, Minneapolis, Minnesota, June 1986.

3. Diaz, L.F., and C. G. Golueke, "Resource Recovery: Imperative for Developing Nations," presented at the 5th International Recycling Congress, Berlin, October 1986.

4. Cal Recovery Systems, Inc., *Composition and Properties of Municipal Solid Waste and Its Components,* report prepared for the U.S. Department of Energy, February 1984.

5. Davies, A.G., "Composting Sewage Sludge with Municipal Refuse," *Compost Science, 1*(3):9, Autumn 1960.

6. Diaz, L.F. and C. G. Golueke, *Composting,* prepared for the Metro Manila Solid Waste Management Study, Republic of the Philippines, May 1982, pp. 132.

7. Senn, C.L., "Role of Composting in Waste Utilization," *Compost Science 15*(4):24-28 Sept.-Oct. 1974.

8. Senn, C.L., *Dairy Waste Management Project: Final Report,* Public Health Foundation of Los Angeles, California, and University of California Agricultural Extension Service, December 1971.

9. Golueke, C.G., D. Lafrenz, B. Chaset, and L.F. Diaz, "Composting Combined Refuse and Sewage Sludge," *Compost Science/Land Utilization,* 21:42.

10. Savage, G.M., and L.F. Diaz, "Solid Waste Processing for the Recovery of Secondary Materials and Energy," presented at the Second Conference on Municipal, Hazardous, and Coal Wastes Management, Miami Beach, Florida, December 1983.

24

SOLID WASTE CHARACTERIZATION

George M. Savage, Luis F. Diaz and Clarence G. Golueke

ONE of the steps most frequently neglected in solid waste management, and yet probably one of the most essential to successful management, is the determination of the size (quantitative) and nature or composition (qualitative) of the municipal solid waste stream—in other words, "characterization" of the waste stream.

Many reasons can be conjectured for the neglect, all of which undoubtedly have some bearing. Three come to mind: 1) Failure to recognize the importance of the step in terms of its essentiality and of its utility; 2) A mistaken assumption that a few counts or visual surveys of truckloads coming to the disposal site can convey all of the information needed; and 3) A conviction that data mentioned in the solid waste literature are sufficiently representative. The last named reason was especially prevalent in the 60's and 70's, but hopefully, has been rather thoroughly discredited in the 80's.

To arrive at numbers and compositional analyses sufficiently realistic to permit valid extrapolations, it is necessary not only to weigh the truckloads, but also to classify and determine the weights and volumes of the various components of the truckloads. With respect to dependence on data reported in the literature, much has been written about the fallacy of such a course. The fallacy stems from the strong relation between the composition and volumes of wastes generated and the locale of generation. It does not require much effort to come to the realization that the volume and composition of solid wastes generated in congested metropolitan areas or industrial cities in the Northeast differ considerably from those of wastes generated in the cities of the Pacific Coast.

Loosely defined, characterization is the determination of mass, volume, and composition, and to some extent, the identification of particular components that could exert an undesirable impact on the public well-being (e.g. hazardous and toxic wastes). Measurements related to characterization extend to rates and totals.

The reasons become especially important when they relate to the development of waste management plans for the present and the near and distant futures, and even more so when resource recovery is concerned. They relate to design, implementation and operation. They have a bearing not only on the concrete aspects of planning but also on the success or failure of the outcome of the plans. Characterization determines not only the size of projections but also their validity.

Waste characterization has taken on new dimensions and implications in that it is being extended to the determination of the danger posed by a particular solid waste stream to the public safety and health. This aspect of characterization has come to attention only relatively recently, and is an outcome of the many reports of environmental contamination that have become commonplace in the news.

Examples:

In this section, a few examples are presented and discussed. One that comes to mind is a situation in which a municipality decides that it must resort to incineration. First, it must determine whether, in reality, incineration would be a useful recourse. If it determines that incineration would indeed be a useful recourse, the next step is to design an appropriate physical plant and attendant operational procedures.

To determine the utility of incineration in terms of its needs, the municipality must know the rate of generation and heat content of its waste stream. To determine the pertinent dimensions of the physical plant and its operation, it is necessary to have a firm knowledge of present generation rates and of the data needed for projecting generation rates to be expected during the life span of the plant.

The economic penalties for errors regarding amounts and rates are severe—and far outweigh the expense of making a waste characterization. With incineration, one is dealing in terms of $50,000 to $100,000/ton of daily capacity. Thus, if a 600 ton/day plant is erected instead of one with the 500 ton/day capacity that turns out to be needed, then it becomes a matter of $10 million of capital needlessly expended. In addition to the over-expenditure, there is the matter of the inefficiency attending the operation of an over-designed plant.

A similar rationale applies to composting and to any other type of resource recovery that may be considered. It even applies to landfill. In the "old days", when land availability was not the problem it is today, projections of life spans of fills were not as critical as they are today. The mounting difficulties imposed not only by the high cost of land, but also by obstacles that must be overcome before the desired or indicated land can be used and the attendant sizeable time element, take away the flexibility that the community once had and demand a precise knowledge of land requirements.

Hazardous Waste Problem:

The awareness of the unfavorable impact had on the public safety by the presence of toxic and hazardous materials in the municipal solid waste stream, is stimulating a public demand that those hazardous substances be identified and quantified. Consequently, studies are being undertaken in which the waste stream is monitored for such substances from the point of generation, through collection, and eventual transfer or disposal. The resulting data are needed for developing plans and procedures whereby required environmental analyses can be made and workers and public health be protected. Equally importantly, they are needed in the development of the hazardous waste management plans that are now being demanded.

Characterization Procedures

Characterization should be accurate, truly representative, and sufficiently inclusive. "Accurate" and "truly representative" are redundant, in that a characterization that is not representative would be lacking in accuracy. However, because of the magnitude of the task and the many difficulties inherent in performing it, a compromise must be made between what would be ideal and that which is realistic. This applies especially to extent and frequency of the sampling, number and nature of laboratory analyses to be performed, and the size of the samples involved in a characterization study.

Sampling:

A source of frustration to those engaged in one way or another in solid waste management, is the lack of a standard consensus test method for the gross sampling and sorting of solid waste. However, the following comments are suggested to individuals who may be delegated to characterize a solid waste output.

Frequency: Inasmuch as the day of the week on which the sampled waste was collected and the season of the year in which the collection took place have a distinct bearing on amounts and composition, sampling should be done on each day of a week and at least once during each season of the year. The frequency can be increased if a greater degree of accuracy is required.

Size of Samples: The ideal size is a truckload, or a representative fraction of a truckload. A number of stud-

Average Waste Composition of Santa Cruz County Packer Truck Wastes (Weight Percent)

	Ben Lomond		Buena Vista		Weighted Average	
Component	**Residential**	**Commercial**	**Residential**	**Commercial**	**Residential**	**Commercial**
Mixed Paper	16.4	26.0	15.7	18.3	15.8	21.1
Newsprint	7.7	5.3	8.9	9.4	8.7	7.9
Corrugated	14.9	17.9	14.5	13.1	14.6	14.8
Plastic	10.4	13.0	9.2	15.5	9.4	14.6
Yard Waste	3.0	1.5	11.4	4.1	9.8	3.2
Food Waste	16.7	6.8	12.9	12.1	13.6	10.2
Wood	0.8	7.7	1.9	4.5	1.7	5.6
Other Organic	5.8	8.3	8.1	4.8	7.7	6.0
Ferrous	6.1	5.4	4.9	5.4	5.1	5.4
Aluminum	1.8	0.9	1.2	1.3	1.3	1.2
Glass	12.6	2.9	9.0	9.9	9.7	7.4
Other Inorganic	3.8	4.3	2.3	1.6	2.6	2.6
Total	100.0	100.0	100.0	100.0	100.0	100.0

ies have addressed the issue of proper size of the sample to be taken from a loaded truck. Unfortunately, information garnered in those studies has not been widely circulated, nor has it been "digested" and converted into test method form. It should be emphasized that the truckloads should be selected such as to be representative of the types of waste and their points of origin. At a minimum, residential, commercial (business), and industrial (roll-off containers) sources of wastes should be sampled. Of late, one or two week periods of sampling seem to have been the norm.

Composition: The extent to which composition should be determined depends in large part on the needs of the planners. Ideally, all components should be identified, classified, and "quantified". The classification poses some difficulties with regard to categorizing certain items. For instance, in which of the three categories, glass, metal, and paper, should an item be placed if it consists of glass, metal, and paper? Additionally, the components can be further subdivided on the basis of grades. For example, glass can be color-sorted. There are more than 100 grades into which paper can be graded.

If incineration were involved, a less detailed categorization could be into combustible and non-combustible fractions, in which case, at a minimum, the heat, moisture, and ash contents of the combustible fraction should be determined. If composting were intended, the categorization could be into biologically decomposable and biologically refractory or inert, and the extent of contamination by hazardous wastes would have to be determined.

Miscellaneous Items:

Additional information to be obtained pertains to that needed for determining rates of generation and total amounts involved as well as the constancy of such data. This latter involves the acquisition of data on economic and population trends. Additional multiple analyses designed to determine the precision of the analytical results should be considered.

Techniques:

Techniques for conducting a characterization study include weighing all wastes brought to the transfer station or landfill during the course of the sampling study, selection of types of wastes and representative truckloads, manual separation of wastes into chosen categories, determination of the net weight and percentage of the individual categories, and collecting and preparing subsamples for other analyses that may be desired (e.g. N,P,K).

25

PREPARING MSW FOR COMPOSTING

John C. Glaub, Luis F. Diaz and George M. Savage

THIS chapter focuses on preprocessing operations — an area where municipal solid waste (MSW) composting involves the employment of substantially different technologies than does the composting of materials such as sewage sludge, manure, or plant wastes.

The large particle sizes of the waste and its composition are two features that render MSW less than satisfactory as a compost substrate upon delivery to the facility by a collection vehicle. Preprocessing operations are often referred to as "front-end processing," which is usually the key to a successful refuse-based resource recovery operation whether the application is composting, combustion, or anaerobic digestion. Indeed, the processing of refuse for composting is subject to the same potential problems that have attended various refuse-derived fuel processing facilities throughout the U.S. Therefore, proper design of the refuse processing equipment and of the processing scheme is crucial to the success of a refuse composting facility.

Because it is a heterogeneous mixture with greatly varying properties, sizes, and shapes, MSW is a complex material in regard to the application of engineering principles. Moreover, data on many material properties that are routinely available in engineering handbooks for common materials (e.g., cast iron, stainless steel, and concrete) are not available for MSW or its component materials.

The design of refuse processing equipment requires accurate information on the quantity, composition, and size distribution of the waste to be processed. Therefore, a thorough waste characterization study is a prerequisite for proper facility design.

Refuse processing considerations are essentially the same for co-composting of refuse and sludge as they are for composting of refuse alone. Therefore, the discussion that follows applies equally to co-composting and refuse composting operations. However, in a co-composting operation, an additional unit process would be added for mixing ("blending") the process refuse and the sludge.

The two major purposes for processing refuse prior to composting are the removal of noncompostable contaminants and the size reduction of the waste materials. Components must be removed that: 1) would pose safety problems to the compost product users (safety issues are discussed here with respect to "macro" contaminants, rather than in terms of chemical or pathogenic contaminants); 2) would interfere with the operation of soil tilling equipment; 3) would interfere with further processing of the organic wastes (e.g., composting); or 4) would detract from the aesthetic characteristics of the compost product. Thus, metals, glass, plastics, and other nonbiodegradable items should be removed.

Different degrees of MSW processing result in different compost product qualities, making it crucial to specify what is required before the processing system is designed. Degree of quality is a function of the uses of the product, i.e., of the markets for the product. Moreover, the relative importance of the various compost properties may differ between users. The key factors that affect quality include level of impurities, stability, nutrient content, particle size, and appearance.

There is a basic trade-off in waste processing between product quality and quantity. In general, for a given processing scheme, as the quality of the product increases the quantity of the product decreases. It also follows that as the quality of the product increases, the amount of reject material increases, and therefore the amount of material requiring landfill disposal also increases. Furthermore, it should be emphasized that the cost of processing is proportional to the degree of processing. Consequently, as product quality increases, the cost of processing also increases.

A high quality compost would most likely be required for such applications as landscaping, home gardening, agriculture, parks, and golf courses. A medium quality compost may be acceptable for land reclamation, final landfill cover, fill material, or subgrade material in landscaping. In some cases, a low quality compost may be adequate for use in land reclamation or as fill material. A low quality compost would usually serve as a suitable daily landfill cover.

Unit Processes

The major unit processes employed in successful mechanical separation systems are discussed in the following sections.

Size Reduction

Size reduction is the single most used and the costliest of the unit processes in refuse processing. The purposes of size reduction processes in compost preparation are: 1) to liberate material components contained within other material components; 2) to provide a particle size suited to efficient operation of separation processes; and 3) to provide a particle size suitable for thorough and efficient composting.

Various types of size reduction equipment are available. (As used in refuse processing, size reduction devices are generally referred to as "shredders".) The most common types of size reduction devices are horizontal hammermills, vertical hammermills, flail mills, pulpers, and shear shredders. Size reduction equipment for refuse can be categorized according to rotational velocity. In low rotational velocity equipment, the primary physical mechanisms by which size reduction is achieved are shearing and tearing. In high rotational velocity equipment, size reduction is achieved by means of impact and tearing.

Among the more important aspects of size reduction are those related to the effect of the size reduction process on other unit processes downstream and on cost. The effect on downstream processes is defined through the size distribution of the shredded product. Cost is largely due to energy consumption and machine wear.

Several parameters affect product size distribution. Among them are feed rate, moisture content of the feed, and configuration of the machine (i.e., grate spacing, cutting elements, operating speed).

An issue that has arisen in the processing of MSW for composting is the practical degree of size reduction. Some people have proposed that MSW can be pulverized into "sand-like particles" (either before or after composting), and as a consequence, that it is unnecessary to remove the inorganic components because they will not be visibly noticeable in the compost product. While it may be technically feasible to size reduce refuse below such a fine particle size, the economic realities of doing so should be well understood before attempting the task.

Energy Use: The energy required to size reduce material increases nonlinearly as the product size decreases.[1,2] Specific energy is defined as the gross energy minus the freewheeling energy divided by the throughput of the material. In order to size reduce refuse to a very fine particle size, very large quantities of energy will be consumed for each unit mass of throughput. The cost impact of producing fine size particles is likewise dramatic.

Machine Wear: In addition to the increase in the rate of energy consumption incurred in size reduction to fine particle sizes, the rate of machine wear also increases.[2] Degree of size reduction is defined as the difference between the characteristic particle size of the feed and the characteristic particle size of the product divided by the characteristic particle size of the feed. The rate of hammer wear per unit mass of throughput increases as the degree of size reduction increases.

Air Classification

Air classification can be used to separate organic and inorganic materials in the processing of refuse to serve as a feedstock for composting. Heavy items in the waste stream, such as glass, metals, rock, leather, rubber, and dense plastics, are separated from the light items. Unfortunately, some yard waste, food waste, and wet paper which are readily compostable items also report to the heavy fraction.

Air classification involves the passing of an air stream through the refuse stream in an enclosed chamber. Usually, the refuse is shredded prior to this state. Separation of particles is achieved as the result of the interplay of drag forces and of gravitational forces. The various air classifier designs can be categorized into three groups, namely vertical, horizontal, and inclined. Vertical air classifiers are by far the most common type employed in refuse processing.

One of the most common positions for an air classifier in refuse processing lines is following primary size reduction. In some cases, a magnetic separation stage is inserted between primary reduction and air classification. Overall, however, successful separation of the organic fraction in refuse has been achieved by processing schemes that employ air classification near the beginning, in the middle or near the end of the processing line. Thus, the positioning is flexible, provided that the air classifier design and operation conditions are matched to the air classifier infeed conditions.

Key parameters that influence the performance of an air classifier include air flowrate, solids (refuse) flowrate, cross-sectional area of the air classifier, column velocity, feedstock, and air density. Of these parameters, operational control usually can be exerted only over the air flowrate and the solids flowrate.

The air-to-solids ratio is defined as the ratio of the air and solids flowrates. This ratio has been demonstrated in numerous field tests to significantly affect the performance of an air classifier.[3] There is a critical air-to-solids ratio above which the light fraction and heavy fraction splits are essentially constant and below which the percentage of material reporting to the light fraction drops off sharply.

Screening

Screening can be employed for the removal of noncompostable materials in the preparation of a feedstock for composting. Materials are separated on the basis of particle size. Thus, characteristic differences in size between certain organic and inorganic materials at a given point in the processing line can be exploited to achieve separation.

Three basic types of screens have been utilized in refuse processing—trommel, disc, and flatbed. Trommel screens are rotating cylindrical screens that are mounted on a slight incline. Disc screens are composed of rotating shafts with discs mounted on the shafts and spaced at specified intervals to provide the desired aperture size. The rotating discs advance the material across the screen. Flatbed screens rely upon vibratory motion to agitate the material and to advance it across the screen surface. Trommel and disc screens have been used successfully in a number of resource recovery plants, whereas severe difficulties have been encountered in the use of flatbed screens.

Screening can be located in a variety of places in the processing line and still accomplish the same general goal, providing that the screen has been properly designed for its particular conditions. Nevertheless, there are also positions in the processing line that are not well-suited for placement of a screening process. Screening can be used to process raw refuse prior to size reduction or to process shredded refuse.

The advantages of screening prior to size reduction are: 1) Glass is removed, thus avoiding the difficulty of removing it by screening after it has been pulverized and impregnated into other materials and 2) The portion of the raw MSW that has a particle size small enough not to warrant size reduction is removed. The amount of MSW requiring size reduction is correspondingly reduced, and the overall cost of size reduction becomes substantially lower. Disadvantages include the heavy-duty construction requirements for the screen and the material handling difficulties encountered when bulky wastes enter the screen. In addition, retrofitting of screens in existing plants prior to size reduction is usually infeasible.

Most applications of screens in resource recovery facilities have been for processing shredded MSW. In some plants, the screen is located following air classification; in others, it is located prior to air classification. Neither approach appears to be superior over the other.

The key parameters that affect the performance of a screening operation can be categorized into construction parameters (e.g., length, diameter, aperture size), operating parameters, and feed characteristics. Typically the only operating parameters over which control can be exerted are the feed rate and the rotational speed of the screen. (With respect to trommel screens, it is the rotational velocity of the screen; and with respect to disc screens, it is the rotational velocity of the shafts.) In some installations, it is possible to adjust the inclination angle of the screen. Feed rate is the dominating parameter in the determination of screening performance.

Magnetic Separation

Magnetic separation removes ferrous metal from the process stream and is one of the most efficient and least costly unit processes used in waste processing. Moreover, the economics of the waste processing operation are improved because the scrap ferrous can be sold.

The two basic types of separators for recovering ferromagnetic materials are the overhead belt magnet and the drum magnet. These pieces of equipment are manufactured either in the form of electromagnets or of permanent magnets. An overhead belt typically is installed above a conveyor that is transporting shredded MSW. In some cases, it is installed parallel to the conveyor, and in others, perpendicular. The magnet is stationary, whereas the belt on the separator normally is set to travel in the same direction as the conveyor belt, but somewhat faster, in order to produce a clean product.

Drum magnets can be installed over a conveyor, over the end of a feeding mechanism, or above a conveyor head pulley. The drum rotates past a stationary magnet. Magnetic material clings to the drum until it is scraped off. Non-magnetic matter simply drops off.

The most important parameters that determine the performance of a magnetic separator are: 1) flux density; 2) rate of change of the flux density (gradient at the object); 3) size of the ferrous object; 4) shape of the ferrous object; and 5) distance between the magnetic separator and the material. Parameters that are interrelated include bulk density, depth of burden, and speed of conveyor belt carrying feedstock to the magnetic separator. In general, using a particular type of magnetic separator and a fixed speed of the conveyor belt carrying the feedstock, the quality or purity of the recovered magnetic metals improves as the depth of the burden decreases.

Processing Schemes

The unit processes described in the preceding section can be employed in a variety of ways to prepare a refuse feedstock for composting. The optimum design of a processing scheme is dependent upon a number of factors, including: 1) the quantity of wastes processed; 2) the composition of wastes processed; 3) the size distribution of wastes processed; 4) the required product characteristics (i.e., market requirements); 5) the available budget for constructing the facility and conducting the operation; and 6) the overall waste management goals of the community.

The preprocessing system comprises the following unit processes:

i) Size Reduction (Shredding): A shredder coarsely shreds all of the incoming refuse.

ii) Magnetic Separation: A magnetic separator removes a large proportion (approximately 80 percent) of the ferrous metal in the waste. This step is not critical to a successful composting operation, because an air classifier could also remove the ferrous. However, the economics of the system are improved because the scrap ferrous can be sold. Moreover, ferrous removal is a simple operation.

iii) Air classification: Air classification removes a large percentage of the inorganics in the waste (e.g., glass, metals, rock). The cyclone is an integral part of the air classification system, serving to remove the entrained solids in the light fraction from the air stream.

iv) Screening: The light fraction from the air classifier contains a sizeable quantity of grit, glass, and small hard plastic particles that can be removed by screening. As in air classification, some organic matter is inevitably removed along with the undesirable inorganic matter.

Prior to composting, the refuse feedstock either is moistened to an optimum moisture content or is mixed with sludge. In addition to the preprocessing system described above, a co-composting operation would necessitate the provision of sludge storage capabilities and equipment for mixing the sludge with the processed refuse.

Recycling of materials such as glass, ferrous, and aluminum by removal at the source of waste generation would be beneficial to the mechanized separation system. The benefits of recovering recyclable materials prior to mechanically processing MSW include reduced wear on processing equipment, reduced contaminants in the compost product, reduced residue going to the landfill from the processing facility, extended landfill life, and the associated economic benefits.

Following composting, the material undergoes further processing to prepare the final product. The composted material is first passed through a device for breaking lumps that may have been formed during composting improving screening efficiency and marketability. Screening provides the desired particle size of the product and removes film plastic and other non-decomposed components from the product. A final size reduction stage may be added following screening, although caution should be taken in attempting to grind to too fine a particle size as discussed in the section entitled, "Unit Processes." Reasonable objectives of a final size reduction stage are to enhance the appearance of the compost product and to improve its spreadability characteristics for field application by agricultural equipment.

A number of other processing schemes also are feasible for preparing a refuse feedstock for composting. For example, a substantially different approach involves the production of both a fuel product and a compost product. Other possibilities include the processing of certain reject streams; however, the increased complexity of such a processing scheme presents added operational risks.

TABLE 1

Allocation of Costs for a Refuse Composting Operation

	Turned Windrow System (%)	Static Windrow System (%)
Preprocessing	40	50
Composting	45	32
Postprocessing	15	18
Total	100	100

Economics

Early in the formulation of a refuse composting project, it is important to define the costs associated with processing the refuse. Indeed, the costs may in turn influence the design of the facility. The initial investment costs of concern are those for the processing building (including the attendant site improvement costs), the processing equipment, start-up and shakedown engineering design and construction supervision, and contingency costs. The major operating costs are those associated with operating labor, maintenance, electrical power, fuel, and residue disposal.

The relative allocation of costs for preprocessing, composting, and post-processing are shown in Table 1 for two composting approaches. These vary as a consequence of different costs associated with the composting stage. Typical unit costs for refuse preprocessing are in the range of approximately $15 to $30 per ton of MSW input. The development of accurate costs requires site-specific information. Moreover, the costs are sensitive to a number of factors, including the scale of the operation, the desired quality of the compost product, and landfill disposal costs. The aforementioned economic data are a consequence of economic studies conducted by the authors using computerized models.[4,5] The models supply both mass balance information on the system and the associated economics.

References

1. Savage, G.M. and G.R. Shiflett. Processing Equipment for Resource Recovery Systems. Volume III. Field Test Evaluation of Shredders. EPA-600/2-80-007c. 1980.
2. Savage, G.M., D.J. Lafrenz, D.B. Jones, and J.C. Glaub. Engineering Design Manual for Solid Waste Size Reduction Equipment. EPA-600/S8-82-028. 1983.
3. Hopkins, V., B.W. Simister and G.M. Savage. Comparative Study of Air Classifiers. Final Report prepared for the U.S. EPA under Contract No. 68-03-2730. 1981.
4. Glaub, J.C., C.G. Golueke, G.M. Savage, L.F. Diaz, D.B. Jones, and D.J. Lafrenz. Feasibility Evaluation of Municipal Solid Waste Composting for Santa Cruz County, California. Final Report. 1983.
5. Glaub, J.C., L.F. Diaz, D.B. Jones and G.M. Savage. Economic Analyses of Large-Scale Composting and Co-Composting. Report prepared for the Metropolitan Council of the Twin Cities Area, St. Paul, Minnesota. 1984.

26

ENERGETICS OF COMPOST PRODUCTION AND UTILIZATION

George M. Savage, Luis F. Diaz and Clarence G. Golueke

THE energetics of MSW composting has many facets, all of which combine to determine the practicality of a compost undertaking. One facet discussed here pertains to the energy cumulatively consumed in the compost process.

Size Reduction: As would be expected because of the characteristics of MSW and the rugged and sizeable machine required, size reduction involves a relatively high consumption of energy, the amount of which is mostly a function of the particle size distribution of the incoming material and of that of the output. For example, taking "characteristic particle size" to be defined as "the size that corresponds to 63.2 percent of the cumulative particles passing through a screen of a certain dimension", about 15 kWh/Mg would be consumed in the size reduction of MSW to a characteristic particle size of 2.44 cm. Among the other factors that have a bearing on energy consumption in size reduction are feed rate; moisture content; and design characteristics of the machine, such as relative speed of rotation of the size reduction devices (hammers, rings, gears, etc.) and the clearance between hammers and grates.

Air Classification: Through air classification, shredded municipal solid waste is separated into a light fraction and a heavy fraction. Among the several types of air classifiers that have been built, the maximum design capacity was approximately 70 Mg/h. With respect to energy consumption, results obtained in a study undertaken by the authors may be taken as an example. In that study, about 4.1 kWh/Mg were consumed in an air classifier system that processed roughly 30 Mg/h. Results obtained by others indicate a range of 3.4 to 4.2 kWh/Mg of throughput.

Screening: The three types of screens most commonly used in the resource recovery industry are the vibrating, the disk, and the trommel. Of the three types, the trommel screen has been found to be particularly attractive because of its relative freedom from operational problems.

Energy consumption in trommeling is approximately 0.8 to 1.1 kWh/Mg of material produced.

Miscellaneous Equipment: Equipment ancillary to or required in the operation of equipment described in the preceding paragraphs include conveyors and mobile equipment such as front-end loaders. Because energy consumption by such equipment varies with a number of factors (e.g., material density, flow rate, size of motor), firm values were not available. However, in the development of overall energy estimates, allowances were made for such energy consumption.

Composting: In the discussion of the energetics of the compost step, it is assumed that the step includes conveyance of the segregated refuse or refuse-sludge mixture to the composting site. The composting step itself is assumed to have a primary ("active") stage and a secondary ("curing") stage. Although there may be some overlap between classes, two classes of systems exist, namely "open" and "enclosed".

Postprocessing: Postprocessing consists mostly of

screening although some size reduction may be required. Screens and equipment are similar to those used in preprocessing.

Summary of Process Energy Consumption

Assumptions made in calculating the energy requirements given in the previous sections were the following:

1) The substrate is a mixture of MSW and digested sludge.
2) The facility can process 1200 Mg of MSW plus 300 Mg of sludge per day (i.e., a total of 1500 Mg/d).
3) The preprocessing plant is designed to recover most of the organic matter in the refuse and to turn out a feedstock from which an acceptable finished compost can be produced.
4) The recovered organic refuse is mixed with the sludge, and the resulting mixture is then composted either by way of a windrow system or by way of an enclosed system.
5) The composting step having been completed, the composted material is further processed to impart to it the desired quality.

The results of the calculations are summarized by the data in Table 1. The values presented for open systems are more firm than those for enclosed systems. The approximately 16 kWh/Mg required for preparing the substrate (refuse) is almost 50 to 70 percent of all the energy used in the compost process as a whole. Depending upon the compost system used, energy consumed in the composting step may be as low as 0.45 or as high as 10.1 kWh/Mg. Postprocessing requires approximately 6.1 kWh/Mg. Miscellaneous functions and activities account for about 1.0 kWh/Mg.

Energy expended in the three major stages, i.e., energy required to produce a metric ton of finished compost, amounts to about 24 kWh with the turned system approach and about 33 kWh with the enclosed system approach.

TABLE 1.

Energy Requirements for Compost Production[a]

Stage		Specific Energy Consumption (kWh/Mg MSW)		
		Windrow		Enclosed
		Turned	Static	Generic
Preprocessing				
Mobile equipment		0.2	0.2	0.
Size reduction		10.0	10.0	10.0
Segregation (air classifier, magnetic separator, trommel, stoner)		4.6	4.6	4.6
Conveying		0.5	0.5	0.5
Mixing (refuse/sludge)		0.7	0.7	0.7
	Subtotal	16.0	16.0	15.8
Composting				
Turner		0.25	—	—
Blowers		—	4.5	—
Aerator, feed, discharge		—	—	10.0
Mobile equipment		0.2	0.1	0.1
	Subtotal	0.45	4.6	10.1
Postprocessing				
Sizing		5.8	5.8	5.8
Conveying		0.1	0.1	0.1
Mobile equipment		0.2	0.2	0.2
	Subtotal	6.1	6.1	6.1
Miscellaneous				
Ventilation, lighting		1.0	1.0	1.0
	TOTAL	23.55	27.7	33.0

[a]*Based on 1200 Mg/d of MSW plus 300 Mg/d sludge.*

27

SOURCE SEPARATION AND HOUSEHOLD HAZARDOUS WASTES

As noted in previous chapters, increasing emphasis on source separation at the household level has a direct effect on improving the trend to solid waste composting. In Europe, the impetus in some cases for source separation has resulted from a goal of eliminating undesirable materials and heavy metals in communities that have solid waste composting facilities. One report notes that 17 communities in West Germany are exploring source separation programs.[1] Similar efforts are reported in The Netherlands and Switzerland.[2,3]

In the United States, the village of Barrington, Illinois undertook a comprehensive curbside/composting program in June, 1988. The Village distributed 90-gallon containers for the collection of compostable materials, which are collected the same day as other recyclables. Early results indicate a 90 percent participation rate.

Another significant trend is the rapid growth in municipal collection programs for household hazardous wastes. The number of programs has grown from only two in 1981 to more than 850 in 1988. Household hazardous wastes include such materials as paints, pesticides, cleaners, solvents, batteries, medicines, pool chemicals, etc.

Reporting on an HHW program in San Bernardino County, California, three directors who developed the special one day collection efforts and permanent collection centers—Richard Roberts, Steve Van Stockum and Diane Christensen—supplied these details of the program logistics:

The waste collected at the permanent collection centers may be stored for up to 90 days in an approved structure prior to transportation to a hazardous waste disposal facility. The following is a description of the type of facilities we use.

Waste materials received are stored securely within a custom built storage shed, which cost approximately $7,000 each. The storage shed is 8′ × 8′ × 22′ overall and divided into two separate compartments, separated by a steel wall. The shed is constructed of 12 and 14 gauge steel with the floor coated with a corrosion-resistant epoxy coating. Spill containment is provided to prevent any unauthorized release should there be an accident. Additional features include:

- ventilation to prevent build-up of fumes;
- portable ramp to facilitate loading and unloading of drums;
- blow-out roof panels to reduce damage in event of explosion;
- fire extinguisher;
- eye-wash;
- portable shower facility;
- placarding to indicate nature of materials stored;
- security locks to prevent unauthorized access; and a
- static ground connection.

The total capacity of the shed is 30 55-gallon drums. The segregated compartments allow for storage of 21 drums on one side and 9 drums on the other. The two compartments are fully segregated to allow for the safe storage of incompatible materials.

All waste materials are accepted only in fully intact containers and with relatively small quantities in each, making the likelihood of a large spill extremely remote. Once accepted the waste materials are categorized and placed in the appropriate side of the shed. Within a few days all materials are lab-packed in 55-gallon drums with vermiculite. Operational procedures are such that the lids are kept on the drums at all times except when materials are actually being placed in them. Drums containing corrosive materials are lined with plastic liners prior to the addition of absorbent and wastes. When 30 drums have been lab-packed and the shed is "full," a contracted hazardous waste hauler removes them from the site. At present, each shed takes approximately two months to fill a total of 60 drums.

It is estimated that over 600,000 gallons of HHW had been entering San Bernardino County landfills each year. Various studies indicate that from 0.14 to 5.0 percent of the household waste stream reaching the landfill is comprised of HHW.

References:

1. Franke, Bernd. "Composting Source Separated Organics," *BioCycle,* 28, 6, 40-41.

2. Lustenhouwer et al. "Source Separation and Collection of Household Compostables," *BioCycle,* 28, 6, 33-36.

3. Mayer et al. "Trends in Yard Waste Composting," *BioCycle,* 29, 6, 60-62.

28

CO-COMPOSTING FOR SMALL COMMUNITIES

Aga S. Razvi

THE need for a low technology, low capital cost composting system that can be adopted by small communities is great. In Wisconsin alone, there are several hundred communities with a population of 2500 or less. Officials of Columbia County requested the University of Wisconsin—Stevens Point to develop such a system. Our small-scale research and demonstration project in 1986 was funded by the State of Wisconsin.

The project site is located outside of Lodi, Wisconsin, about 20 miles north of Madison. A 75 foot by 300 foot building was used which has a concrete floor and additional paved space adjacent to the building. The building was not heated during the winter.

The solid waste is shredded and sewage sludge is mixed with it. The mixture is placed in a static pile which is aerated. To make a finished compost, the material is screened and put in a curing pile. Screened coarse rejects are taken to a landfill.

Solid wastes are taken on a need basis. Normally two truckloads of waste (8-10 tons), representing about 1500 people for a rural community are taken in a week. Two communities with active source separation programs provided their curbside waste. Waste characterization studies have shown that compostable materials make up 73 to 85 percent of the waste. Source separation programs in the community could change this number slightly. Regardless of the change, source separation programs that separate glass, metal, newsprint and cardboard would be complementary to composting and should be strongly encouraged.

The composting system is very simple and utilizes off-the-shelf equipment. One full-time operator and a part-time helper provided the labor. Solid waste is fed into a hopper which feeds onto a horizontal sorting conveyor (this is only 12 feet long), and from there to an elevating conveyor into the shredder. Another conveyor takes it into the mixer.

The garbage bags are opened before shredding, primarily for inspection, because the shredder has a very small size throat. The size of the shredder (15 hp) was determined by the power supply (single phase) at the site. The shredder throat was 14″ × 27″, not large enough for an entire garbage bag.

Many of the non-compostable items were hand sorted and removed to a landfill. These included clothing, hard plastics, leather and rubber, metal, glass, etc. A portion of these materials could not be sorted; therefore they were shredded and became part of the compost pile. Some compostable materials such as large cardboard boxes were also removed. Overall, an average of 16.4 percent of the incoming waste is sent to the landfill.

Shredding and Mixing

The shredder has slow-rotating knives that are mounted on two shafts. The knives are one-and-a-quarter-inch wide. Approximately three-quarter tons an hour of municipal solid waste is processed in this small shredder. The shredded solid waste was put through the shredder a second time to further reduce particle size.

The material is then put into a farm-feed batch-operated mixer that has three horizontal augers. Off-the-shelf equipment was chosen to show towns that they didn't have to devise new equipment. When the mixer hopper is full, its contents are weighed by a tension scale. An inoculum made of medium screenings from a previous compost pile is added to the solid waste. The inoculum was approximately 20 percent by weight of the shredded solid waste.

Anaerobically digested sludge was added as a liquid source to the solid waste-inoculum mix. The sludge had 2 percent solids. Sludge contributed very little nitrogen to the mixture. Water can be used instead of the sludge to achieve the same result. The mix is 40 percent sludge to 100 percent by weight of solid waste mix. The final ratio would be about 50 percent moisture content. If a community has available a dewatered sludge, a larger percentage of sludge would have to be added to achieve a 50 percent moisture.

The sludge-solid waste-inoculum is mixed for about 20 minutes, which results in good pulping action and some particle size reduction. Approximately 90 percent of the particles are smaller than one-and-one-quarter-inch at this point. The mixed material is ejected from the mixer and placed in a pile enclosed by a snow fence.

Pile Dimensions

The piles were 12 feet in diameter. The snow fence was four feet high, so the pile was roughly four feet high at the perimeter and coned to about six feet. Two truck loads of solid waste (8-10 tons) after processing would fit into a pile of this size. Some piles were constructed in the shape of a windrow and contained waste from 5-7 truckloads.

This demonstration project experimented with different variables. In order to compare variables, piles had to be constructed within short time (weekly) intervals, because of the climatic changes. Generally all waste from one week was placed in one pile.

Temperature was monitored at different depths in each pile on a regular basis. Temperatures reached 132°F within 3 days after pile construction during the summer and within 11 days during the winter. (During winter, the refuse is frozen when placed on the curb the night before pickup.)

Oxygen was monitored at different depths on a regular basis. All piles were aerated either with positive or negative pressure aeration. Negative pressure aeration works better than positive pressure aeration, especially for Wisconsin winters where ambient air temperatures could be below zero for prolonged periods. When air this cold is blown into the heart of the pile, the pile rapidly loses heat and pile temperatures fall below ideal. An attempt was made to maintain at least a 3 percent oxygen level in the piles at all times. This resulted in an odor-free operation.

Prevention of Problems

No free liquids were produced from the compost piles. This eliminated possible breeding places for flies and mosquitoes. Although flies were brought in along with the refuse, covering the newly constructed pile with a thin layer of finished compost seemed to eliminate flies. No additional water was added to the pile after the initial mixing with liquids.

In an attempt to determine the amount of leachate produced from rain and snow, some piles were placed outside. All precipitation falling on the pile was immediately absorbed at the surface resulting in no downward movement of free liquids to produce leachate.

Piles placed outside of the building showed no signs of blowing paper or plastics due to wind. The fairly large particle size resulted in the entanglement of particles during mixing/pile construction. Formation of a dry crust at the surface further prevented blowing. When the surface of the pile was wet, the particles were too heavy to be blown by wind.

The compost is ready to be screened in six to eight weeks after pile construction. By this time, the pile height is about 2/3 to 3/4 of its original height due to decomposition and settling. In preparation for screening, the pile is partially dried. Continuous positive pressure aeration is provided until the moisture content decreases to about 30 percent. The pile is now ready to be screened.

Screening Material

The pile is broken up and taken out for screening in a double drum grain screen, again available from a farm equipment dealer. The material is fed into the inside drum, which has a coarse screen on it. The outer drum has a fine screen. The material then drops into three portions, according to particle size. Each fraction is weighed to determine mass balances.

The coarse screen has a one-inch opening. The material retained on the coarse screen is primarily made up of film plastics. Metals make up 10 percent by weight of this fraction. About one quarter of this material is compostable. Coarse screenings account for 16 percent by weight of solid waste received at the facility and is landfilled.

The final product, is one-quarter inch screened material that is rather fine, is mixed with water and allowed to cure. Curing pile temperatures will reach 132°F within a few days. Curing will further ensure pathogen destruction as well as improve the appearance of the product to a dark, crumbly material.

All compost produced at the site is being stored until greenhouse and field testing can be completed. The compost has been tested for heavy metals, plant nutrients and priority pollutants. Nitrogen content of the finished compost ranges from 1.4 to 2.5 percent. Phosphorus at 0.58 percent and available potassium is at 0.59 percent.

On a weight basis, the final compost makes up about 21 percent of the weight of the incoming waste. This fraction will depend on the composition of the incoming waste and the time allowed for composting and curing. There was a 28 percent weight loss due to decomposition of the organic materials in the solid waste.

Medium-sized screenings, the fraction larger than one-quarter inch and smaller than one inch, accounted for 19 percent by weight of the incoming solid waste. Glass, metal and plastic make up 28 percent by weight of this fraction while the remaining 72 percent is compostable. Because of the large compostable fraction, the medium

FIGURE 1.

Mass balance for co-composted municipal solid waste.

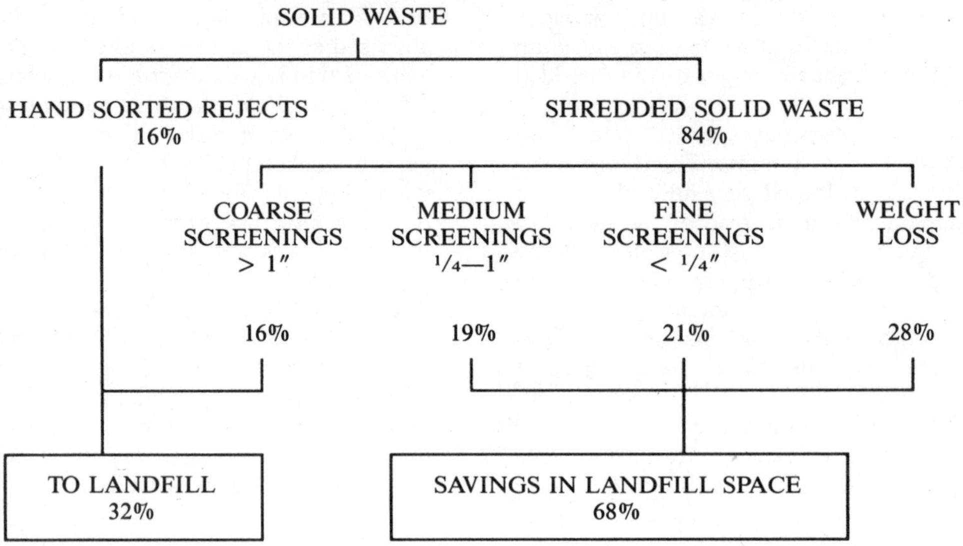

screenings are mixed with uncomposted shredded solid waste as an inoculum. The active microbial population in this fraction promotes rapid temperature increase in a newly constructed pile. In addition, the compostable portion of the medium screenings will undergo further decomposition in the new pile.

The overall mass balance for three-month-old piles is presented in Figure 1. These numbers would vary depending on the composition of the incoming solid waste, the solids content of the sludge, the pile detention time, the intensity of source separation/recycling in the community, size of screen used, etc. The results of this composting study suggest that there could be a 68 percent savings in landfill space. Only the hand sorted rejects and the coarse screenings need to be landfilled. Landfill savings could be even greater with active source separation/recycling programs.

Financially, the system could be set up (as of 1986 calculations) to process 10 tons a day, five days a week for under $200,000 for equipment costs. With typical capital, operation and maintenance costs, tipping fee of less than $40 per ton can be expected. This does not include any revenues for the compost.

This system will work for a small community that wants to be independent of a large landfill, either because of the distance or for political reasons. The low capital cost allows for distributed rather than centralized solid waste processing facilities for communities in rural counties.

Dr. Razvi is on the faculty of the College of Natural Resources, University of Wisconsin—Stevens Point.

SECTION SEVEN

LEAF AND YARD WASTES

29

YARD WASTE UTILIZATION

Jim Glenn

As disposal capacity for solid waste diminishes, many program managers spend considerable amounts of money designing and implementing curbside recycling programs to divert newspaper, glass, metal containers and other materials for the municipal waste stream. More and more, however, they are looking at what can be the easiest and most cost effective approach to diverting waste from other disposal facilities—yard waste utilization.

As one would expect, the amount of yard waste available in any community varies widely from one part of the country to another. Most waste composition studies put the percentage at anywhere from 10 to 30 percent of the municipal solid waste (MSW) stream. Franklin Associates, in a waste characterization study done for the U.S. Environmental Protection Agency, found the total amount generated in the U.S. in 1984 to be slightly under 24 million tons annually, or 18 percent of the total amount of MSW generated. By comparison, that is almost one half of *all* paper and paperboard products discarded in 1984.

The variation in the amount of yard waste generated throughout the year is dramatic. Grass and garden waste is usually generated in the greatest volumes from late spring to early fall. For instance, Omaha, Nebraska estimates that approximately 50 percent of the residential waste picked up by its crews during the summer is grass. Tree and shrub prunings are usually generated in the spring and fall. Most leaves are collected in the fall, with a significant but lesser amount generated in the spring. The fall leaf drop places such a burden on the MSW collection system that many communities collect them separately. It is not unusual for leaves in some communities to represent 15 to 20 percent of the total solid waste stream during the entire year, and 60 to 80 percent of the waste generated in the fall.

In addition to the significant diversion potential, yard waste utilization projects can be developed in a very economic manner. Communities that already separately collect yard waste simply have to redirect collection vehicles from a landfill or waste-to-energy facility to the utilization program. Most yard waste projects include a composting facility to process the material. Another approach, depending upon location, is to use the yard waste directly, e.g. applying leaves to nearby farmers' fields. For example, Lower Swatara Township in Pennsylvania provides its leaves to farmers, who use the material for animal bedding. Many communities used chipped woody material for mulch and park trails.

Another very decided advantage to yard waste utilization is the market situation. In recycling programs, most of the material collected is sold on the open market and thus must compete with materials from other programs—and ride the ups and downs of external markets. The majority of the current yard waste projects, on the other hand, operate in a closed loop system. The material generated is utilized within the community from where it came. In most cases, either the municipality uses it for landscaping purposes or cover at the local landfill, or it is given back to citizens for their use.

Legislative Initiatives

Because yard waste can be diverted and put to use so easily, literally hundreds of communities throughout the U.S. have developed utilization projects. In the upper Midwest (Michigan, Wisconsin and Minnesota), there are more than 200 known projects. A recent survey in Pennsylvania identified more than 75 projects. But nowhere else has yard waste utilization escalated more rapidly than in New Jersey, where more than 180 leaf composting sites have been permitted, with many more to come on line in the next year.

Success at individual community yard waste utilization projects has led a number of states to formulate strategies aimed at encouraging the development of even more programs to help alleviate disposal problems. The efforts have taken a number of forms. The most dramatic is prohibiting the disposal of yard waste in solid waste disposal facilities other than composting sites. Minnesota and Wisconsin recently passed legislation that would ban the disposal of yard waste, including leaves, grass clippings, garden debris and brush, from disposal facilities before 1993. New Jersey's ban, which took effect in April 1988, only covers leaves and not other types of yard waste.

Pennsylvania has proposed a ban on leaves that would be enforced two years after the effective date of the legislation. The long lead time in implementing these bans (in Minnesota and Wisconsin and to a lesser extent in Pennsylvania) is due in part to the need to develop utilization systems to accommodate the yard waste. In New Jersey, however, the system needed to replace landfilling of leaves is already largely in place.

The prohibition mandates have led a number of county governments in those states to restrict disposal of yard waste prior to the date set by the state legislation. In Wisconsin, both Dane and Eau Claire Counties have passed ordinances restricting the disposal of yard waste and Sauk and Juneau Counties are considering similar measures. Broome County, New York banned leaves from its landfill in 1986 after developing an alternative composting facility at the County's landfill site.

A less drastic approach used to encourage yard waste composting is to allow those facilities to be permitted and operate under less stringent regulations than ones that apply to other types of solid waste processing facilities. Most states that take a less stringent approach do so because they view yard waste as a resource rather than a waste, in much the same manner that source separated paper, glass and metal are viewed as a resource. Many state regulatory agencies also believe that the potential for environmental problems at yard waste composting facilities is much lower than at mixed waste composting facilities and other waste processing systems.

States have devised a number of strategies in their efforts to allow yard waste composting facilities to operate under less restrictive guidelines. For instance, in Michigan, yard waste that is source separated and utilized is not considered a solid waste and thus not regulated by the Department of Natural Resources. While the solid waste laws of the state do not apply if the facility violates regulations concerning air emissions, odors, groundwater contamination, etc., appropriate action can be taken under those statutes.

In Wisconsin, the degree of regulation of yard waste composting projects varies with the size of the project. Systems sized to handle less than 50 cubic yards of material annually are not regulated. Projects handling 50 cubic yards to 20,000 cubic yards are governed by a set of guidelines based primarily on site considerations and performance standards. Projects over 20,000 cubic yards in size are subject to a formal permitting process. Pennsylvania recently passed solid waste management regulations that exempt leaf composting sites from regulation if a facility follows the guidelines established by the Department of Environmental Resources.

New Jersey retains perhaps the most control over projects without requiring a complex permitting procedure. Its Department of Environmental Resources requires that all leaf composting projects be permitted regardless of size, but the permitting requirements are considerably streamlined in comparison to the procedures for other solid waste processing projects. Even though leaf composting projects do not require sophisticated modeling and engineering drawings as part of the permitting procedures, it is not unusual for the process to take a year or more to complete.

Yard waste-based products are usually marketed or utilized in the community where they are generated. To help ensure that additional markets are available, several states have begun to give a preference to compost when purchasing material for land maintenance. In 1985, Minnesota's governor established a preferential purchasing program, noting at the time that "State agencies can lead the way in showing how wastes can be used to improve land instead of destroying it." The mandatory recycling law passed by New Jersey in 1987 has provisions that require "all State and local agencies responsible for the maintenance of public lands" to give "consideration and preference" to compost in all land maintenance activities.

These purchasing preferences have encouraged state agencies to develop specifications for compost products. As would be expected, Minnesota state agencies were among the first to complete work in this area. The state's Department of Transportation developed specifications that will be used in the purchase of compost, while the Department of Agriculture has completed work on ones that relate to consumer protection. The Minnesota Pollution Control Agency has developed rules governing the use of compost. Additionally, New Jersey's Office of Recycling and Massachusetts' Department of Environmental Quality Engineering (DEQE) are working with their respective state purchasing agencies to develop appropriate specifications for compost products.

30

TECHNOLOGIES FOR LEAF COMPOSTING

As part of their efforts to aid communities to adopt sound methods for composting leaves, members of the Rutgers University faculty, state extension service and New Jersey Office of Recycling developed data on various methods using different levels of technology. The technologies — described by P.F. Strom, F.B. Flower, M.H.P. Liu and M.S. Finstein — originally included minimal, low-level and high-level with an intermediate method subsequently advanced. The technologies described are as follows:

1) Minimal Technology: Leaves are formed immediately into large windrows, e.g. 12 ft. high by 24 ft. wide using a front-end loader. Once each year the windrow is turned and reformed. A new windrow is constructed each fall with the new leaves. After three years or sometimes longer, the material in the windrow is stabilized and ready to be used as compost.

The minimal technology is not designed for rapid composting. Much of the pile remains anaerobic for a full year between turnings. Unpleasant odors can be expected prior to the first turning, and serious odors may be released after the first turn. Therefore a leaf composting site using minimal technology should be relatively isolated. The advantage of this approach is that it is extremely inexpensive, with relatively little space required for the actual composting of the leaves. A usable compost is produced, albeit with the potential for serious odor problems during turning.

2) Low-Level Technology: A low-level method achieves more rapid composting of the leaves under conditions that would not generate offensive odors. In particular, this means having to ensure adequate conditions of moisture content, oxygenation, and temperature within the pile.

Prior to or during initial windrow formation, a minimum moisture content of 50 percent is desired (a rough approximation is about 20 gallons of water to be added per cubic yard of collected leaves). The desired temperature range for composting is 70°F to 140°F, which can be achieved by adjusting pile size. But oxygenation levels can be adversely affected by pile height. To balance the two, the recommendation is to start with moderate size piles — 6 ft. high by 12 to 14 ft. wide — then combining two piles after the first burst of microbial activity, which lasts approximately one month.

The piles should be turned in the spring to increase the rate of decomposition and improve destruction of any pathogens or weed seeds. During turning, maximum mixing and "fluffing" is desired. Water can be added if the material is too dry, but this is not necessary if the material was wet enough initially. Additional turnings throughout the summer will further enhance the composting rate and product quality. While much of the material will not be stabilized completely by the end of the summer, the authors point out that the material is fairly well decomposed, has low oxygen demand, and is unlikely to generate unpleasant odors. This material can be put into larger curing piles built around the perimeter of the site. While the piles can be made as large as desired to conserve space, compaction should be avoided. A well stabilized, finished product should be available by the following spring. The authors add that shredding is an optional step, one which improves the physical quality and appearance of the finished compost, making it more acceptable to the public for many uses. The disadvantages to shredding are increased capital and labor costs, as well as a need to dispose of rejects.

3) High-Level Technology: Space constraints and a desire to complete composting in a year led to the development of the high level of technology. It involves forced aeration of the large windrows with blowers. (The authors point out that turning the windrows more frequently with a front-end loader can produce a finished product within a year, however, since the original windrows can be no larger than those of the low-level technology to avoid odor problems, little space is saved.) The blowers are controlled by a temperature feedback system to ensure near optimum temperatures in the bulk of the material while maintaining a well-oxygenated condition at the same time. Nitrogen (fertilizer) could be added to further speed decomposition. Adequate pre-wetting of the leaves is a necessity, and after perhaps one month of high rate processing, the blowers could be removed and mechanical turning of the piles would begin. Although the high-level technology approach speeds decomposition, the authors recommend a moderate size buffer zone because the incoming leaves may themselves be odorous and these odors may be released during the initial windrow formation and start-up.

Technology Update

Dr. Peter Strom of Rutgers University notes these recent developments in leaf composting research: "We have added one more level of technology," says Strom, "an intermediate level that includes the use of a windrow turning machine. But in New Jersey, the low-level technology approach is the most prevalent, although some of the regional and commercial sites are going to the intermediate approach. They do weekly turnings with a turning machine and have a finished product in four to six months. Because they are using the machines, generally the intermediate technology facilities actually need more than the 1 acre per 3,000 yards of leaves that we recommend because the pile height has to be five feet or less to accommodate a turning machine. The oversize Scarab, however, can handle higher windrows. The machines give more efficient and better turning, but of course there is also the increased capital cost."

31

IMPLEMENTING A COUNTY-WIDE PROGRAM

Brian R. Golob

THE process by which Ramsey County, Minnesota implemented a leaf composting program illustrates the steps needed for success. Ramsey is the smallest, most densely populated county in Minnesota, with Saint Paul comprising the largest geographic portion and 60 percent of the population.

Yard waste represents approximately 9 percent by weight of the total MSW and 19 percent of the residential waste stream. Of the 9 percent, fallen leaves comprise approximately 2 percent by weight, grass clippings 6 percent, and brush and garden wastes the remaining 1 percent.

Ramsey County staff identified a 7-step planning process for implementing a network of yard waste composting sites in the early 1980s: 1) Identify potential markets, 2) Estimate quantity of yard waste generated, 3) Estimate land requirements, 4) Site selection, 5) Site preparation, 6) Project publicity, and 7) Site operation.

The first step involved identifying markets for the compost. Staff established a ranking of end user markets at the onset of the program. It was assumed that a high quality leaf compost that was thoroughly decomposed and resembled soil would be readily accepted and used by different end users.

The general public was identified as the primary end-user for the compost. The County would allow residents to take compost on a first-come, first-served basis. If the supply of compost exceeded the citizen demand for the material, the County's Public Works and Park and Recreation Departments would be contacted and encouraged to use the compost. If the Departments could not use the compost, staff would contact local landscapers and nurseries to see if they wanted the material. As a last resort, the material would be spread around at the compost sites to increase the organic content of the soil.

Staff estimated the quantity of yard waste, especially grass clippings and leaves, which was being landfilled. Several assumptions were used to conservatively estimate the amount of yard waste generated in a municipality during one calendar year: the average number of bags of leaves per home, the average dry weight per bag of leaves, the average weight for a bag of grass clippings, the average number of bags of grass clippings set out for collection each week, the length of the growing season, the number of single family dwelling units, and the percent of single family dwelling units serviced by refuse haulers.

The number of single family dwelling units was used rather than total occupied households, because there were no reliable estimates for the amount of yard waste generated from multi-family residences, or commercial and industrial properties. Consequently, the yard waste quantity estimate was probably low.

Estimate Land Requirements

The amount of land required for a composting operation depends on the volume and type of waste which will be composted, and the type of equipment used for processing the material. The size of a windrow and the spacing between each one depends on factors such as: the geographic location of the community, the site dimensions, and the equipment used to aerate the windrows.

One acre of land is equivalent to a parcel roughly 208 feet by 208 feet. County staff estimated that one acre was adequate for approximately 3,000 cubic yards of material, or three 1,000 cubic yard windrows based on the size and geometry of a windrow. In Minnesota, large windrows were needed so the piles would not freeze during the cold winter months. The outer 12 to 18 inches of the pile functioned as an insulating blanket.

Recommended windrow dimensions range from 20 to 25 feet across at the base and 10 to 12 feet high. In practice, the windrows' height was limited by the size of the front-end loader. The length of a windrow depends on the size of a site. In Ramsey County's program, the windrows were spaced approximately 40 to 42 feet apart from one another which provided sufficient room to turn or roll them, and allowed for a front-end loader to manuever between piles.

Site Selection

Most cities or counties have a limited number of potential areas from which to select compost sites. This problem was especially evident in Ramsey County. These site selection factors were carefully evaluated: location, size, accessibility, surface water drainage patterns, proximity to nearby homes, and the presence or absence of a visual screen.

Identifying and selecting potential compost sites was a difficult task because there were very few vacant, undeveloped parcels of land in the County. The first step in the selection process required staff to select target areas for compost sites based on their geographic location within the County. It was very important for the target areas to be uniformly distributed throughout the County, with the goal of maximizing citizen participation.

After selecting the target areas, staff contacted City Managers in the surburban municipalities and neighborhood Community Organizers in the City of Saint Paul to assess their interest, obtain support, and identify a contact person. Staff requested each contact person identify vacant parcels in their area which might serve as compost sites. Environmental Health staff then inspected each potential site identified. After examining all of the sites, staff decided that one site in each target area be developed into a compost site.

Staff wanted the municipalities and neighborhood groups to develop a sense of ownership and pride in the program. Several city staff members in some municipalities were leery of participating in the new program. But their initial skepticism turned into enthusiasm after witnessing hundreds of people using the site, and the fact that County staff had done a good job of coordinating the new program.

Site Preparation

Ramsey County Public Works Department staff developed each site. Site preparation included: mowing onsite grass or weeds; trimming tree branches; dozing and grading an access road; grading the site; hauling and spreading road base material to support traffic; and installing fences, gates, and signs.

Easy access to and from the site was important. A two lane access road was constructed from recrushed asphalt or gravel at the sites. The composting area was covered with gravel at two sites, and by grass at the remaining sites. Gravel was used at two sites because Public Works staff decided the ground was too soft to support vehicles. Staff did not want anyone getting their car stuck in the mud; unfortunately, several cars got stuck in soft ground at a couple of sites after it had rained and before the sites dried out.

One problem resulted from using gravel to cover an entire site. The following spring, a front-end loader operator inadvertently mixed some gravel in with the leaves as the pile was aerated, contaminating the compost.

Several metal road signs were installed at each site. For example, two or more 24″ by 36″ directional signs (depending on the configuration of nearby roads), one 18″ by 24″ site identification sign, and several 12″ × 18″ informational signs (e.g., hours and rules signs). A simple gate consisting of two guard rail beam posts, steel cable, cable clamp, master padlock, and a plastic reflector was installed at each site to control illegal access when the site was closed.

Publicity

Publicity is an essential element when a program depends on public participation. Since the concept of centralized leaf composting was new to most County residents (during the fall of 1983), a general publicity campaign was used to introduce the concept. The principal objective of the campaign was to persuade homeowners to bring their leaves to a site. This was accomplished by explaining what composting was, how to do it, why it was done, and where more information could be obtained.

Project publicity efforts were divided into two categories. First, information was developed which promoted the concept of composting and the County program. Second, specific information was developed for each suburb or neighborhood district.

The first task of the publicity campaign involved adopting a slogan. The purpose of a slogan was to reinforce a message in the public's mind. The County's theme was and still is "Lend Us Your Leaves." Program publicity included a variety of methods.

The principal method of publicizing each site to local residents was a two-page brochure delivered to each home in a designated area. A different brochure was developed for each of the sites. A total of 38,500 brochures were printed and distributed. The distribution method varied with each site. Some brochures were bulk mailed, others were inserted and mailed along with local city newsletters, but most were hand delivered by teenagers from scout organizations and swim clubs.

Brian Golob had been project manager of the Ramsey County Leaf Composting Program and is presently on the staff of Pope-Reid Associates in St. Paul.

32

LESSONS FROM OPERATING PROJECTS

THE following summaries of demonstration projects and operating facilities provide insights into collection programs, privatization developments and operational techniques:

Biodegradable Bags: Lincoln, Nebraska has established a yard waste collection and composting demonstration project using cornstarch biodegradable additive for polyethylene bags. According to Gene Hanlon and Gary Brandt of the Lincoln public works department, the city chose the bags to eliminate the costly process of separating plastic bags from the compost pile.

Prior to the start of the project, a survey was mailed to each household to determine the number of households having yard waste and their support in using cornstarch biodegradable bags. There was a 39 percent response to the survey. Roughly 74 percent of responding households leave yard waste for the garbage haulers to collect sometime during the growing season. Twelve percent of the respondents have a lawn service haul away their grass while an equal amount mulch or compost their yard waste. A little over one percent of the homeowners do not collect their grass clippings when they mow.

In addition to collecting yard waste from homes in the demonstration area, all yard waste taken to the landfill by small vehicles and lawn service companies is diverted to the composting site. Over 280 tons of material was collected during the month of June, 1988. The city rented equipment to shred the cornstarch bags and yard waste it has received. A portion of this material will be mixed with sewage sludge so that the city will have some finished compost by the spring of next year. In addition to yard waste and sewage sludge, the city plans to experiment mixing various portions of pen waste from Nebraska's State Fair Park, wood chips and leaves.

The city would like to keep its composting costs low; thus, it is utilizing a relatively "low-tech" approach to composting. The windrows containing sewage sludge will be turned at least twice a week while the turning frequency for the windrows containing grass and leaves will depend upon the pile temperature.

The cornstarch bags have a 33 gallon capacity.

Traditional plastic takes anywhere from 4 to 400 years to decompose. Promoters of the cornstarch bags say that they will degrade within 18 months or 6 years depending on how it is handled. The bags utilized in the Lincoln project contain a cornstarch biodegradable additive amounting to 6 percent of the bags' composition. In addition, the bags contain varying amounts of a photodegradable chemical and a pro-oxidant chemical which make the bag more susceptible to oxygen and heat decomposition. In an active compost pile, it is estimated that these bags will degrade within the 18 month period.

Mechanical Turning: At a regional composting facility in Morristown, New Jersey, more than 27,000 cubic yards of leaves from 7 Morris County municipalities are being composted with use of a large mechanical turner to turn the windrows. Grass clippings are added to the windrows. Four more facilities are planned for the county, according to Alan Little who supervises the program. To haul the machine (a Wildcat compost turner) from site to site, a conventional trailer was converted into a lowboy by removing the sides, installing permanent tie downs on it, and changing the hitch system so it can be pulled by a dump truck or loader. "Essentially composting becomes a one-man operation," says Little.

Morris County advanced the money for yard waste composting and the municipalities would gradually pay it back at a charge of $4.10 a cu. yd. "That may seem like a high price, but everyone was satisfied. The alternative was a cost of $110 a ton to haul to a transfer station," points out Little. "There's no profit in it for the county, just a breakeven rate based on land and machinery costs and the average employee salary, and enough more so that at the end of a 10 year cycle we can purchase new equipment." Finished compost is delivered back to residents at $40 per seven yard load. "That pays for the hauling," says Little.

Process-wise, the Morris County facility has found that five feet high and about 10 feet wide is the right size windrow for turning with the Wildcat. "But we had to build larger piles to accommodate all the leaves coming in last fall," says Little. "These larger piles hold the moisture better anyway. Later, after they had shrunk a little, we repiled them with a front-end loader."

Claw and Grind Operation: In 1981, the Davis (California) Waste Removal Company initiated a composting program to process much of the plant wastes that had been hauled to the landfill. Processing includes removal of contaminants from the collected material, grinding the vegetation in a tub grinder, windrowing, composting and eventually marketing.

Efficient collection is based on use of a pincher bucket attachment originally developed by Sam and Bill Hart of Davis. Residents are asked to keep yard waste separate from other household wastes and place it in the street for separate collection. A two-person crew, one in the loader (with claw attached) and the other in a rear-loading packer truck, are able to collect over 400 tons of clean yard and garden prunings per month.

Efficiency in the Windrow: Part of the leaf composting program in Wellesley, Massachusetts is done in the Department of Public Works, 80-acre yard. There are four large windrows, 10 to 12 feet high and over 100 feet long, explains Pat Berdan, Public Works Director, who supplies this data:

Leaves are composted in the windrows for about a year—from the time the leaves come in one fall season until we consolidate them into a curing pile the following October. A calculation of the material in the four windrows, taken at the end of August, 1987, indicated a total of 2,552 cubic yards which came to 1,387 tons. Snow fence has been installed around the pile to keep the compost from "walking away." Leaves are trucked in each fall by commercial landscapers to whom we sell permits at $150 per vehicle for the leaf season, or until the yard gets filled, whichever is first.

After the leaves are pushed into the curing pile, we screen the material with a Royer or Reed Screen-All. After screening, this material is used for maintaining municipal lands, such as parks, recreational facilities, etc.

Commercial Operations: Increasingly, privately-run facilities are composting yard wastes. Outside Portland, Oregon, Rod Grimm has developed a program to receive loads of yard waste from private haulers and landscapers, and eventually marketing the composted product. On the East Coast, firms like Recycled Wood Products, Compost Management and Woodhue Farms charge a tipping fee lower than alternative disposal methods and compost a wide range of materials. Woodhue, for example, produces a soil conditioner from food processing waste — cranberries, squeezings, leaves and stems. At a 47-acre site in East Moriches, New York, Metski Enterprises, a mixture of trees, stumps and duck manure is composted with leaves. The company processes between 100,000 and 125,000 cubic yards a year, running them through a low-speed shredder before windrowing.

SECTION EIGHT

INDUSTRIAL AND HAZARDOUS WASTES

33

USING INDUSTRIAL WASTES IN AGRICULTURE

ALMOST 40 years ago, Rutgers University soils professor Steven Toth compiled a list of organic waste materials of value in crop production. "It is surprising to find that in a state as highly industrialized as is New Jersey," he wrote, "that stores of organic residues are available in amounts sufficient to add one ton of organic matter to all the cultivated land in the state. The supplies are roughly 1½ to 2 million tons." The list compiled by Toth began with apple pomace, brewery and cannery wastes, went on to fish products and leather dust, and ended with whey and wool wastes.

Since 1960, reports have been published regularly on the composting of industrial wastes on that list. For example, Cornell University agricultural engineers found that "the pharmaceutical industry can be one important source of composted organic residuals." According to Lewis Naylor and James Johnson, recycling mycelium residues to land via composting allows the material to become an agricultural resource rather than an industrial waste.

Composted mycelium residues and composted wastewater treatment sludge were obtained from Lederle Laboratories, Inc., Pearl River, NY for use in their research. These two composts were used individually and combined 1:1 in the experimental work. The dewatered mycelium residues were composted using the static pile forced aeration method. The sludge from wastewater treatment was dewatered, mixed with leaves from streetside pickup, and windrow composted. The two composts were shredded and combined 1:1 before experimental use.

The physical and chemical properties of these composts indicated potential for their use as turfgrass amendments.

Similar results have been reported with paper mill sludges. In Portugal, paper mill sludge was mixed with woodchips as a bulking agent. Researchers at the Portuguese Cellulose & Paper Company in Aveiro reported that the heavy metal content in the original sludge and in the obtained composts was below the limit values specified by EEC directives. "The end product, besides the advantages of a normal humic corrective, has also value for acid soil correction because of the calcium carbonate conditions," they found.

Fishery byproducts represent 15 percent of mid-coastal Maine's solid waste, exceeding 8,000 tons annually. Fish scraps and crab shells were mixed with sawdust and/or woodchips in varying amounts, and windrowed under the eye of consultant Will Brinton, owner and director of Wood's End Labs in Mt. Vernon, Maine. "The pilot study was successful," says Brinton, "and has stimulated state and business leaders into beginning plans for more composting on a commercial level."

The various mixtures composted sometimes displayed interesting differences. For example, windrows where bio-ash was mixed into the sawdust/fish waste reached higher temperatures during composting, and decomposed nearly twice as fast. Nutrient content (NPK) of the finished compost was low, as expected. Bill Seekins of the Maine State Department of Agriculture has done a rather complete inventory of compostable wastes. The list is long and varied: papermill sludge (the highest in quantity produced in Maine), apple pomace, wood byproducts, wood ash, potato processing wastes, kiln cement dust, animal manure, seafood wastes, and other more esoteric materials.

Woolen Fibers

Brinton has an interesting woolen mill sludge project under way with the Robinson Manufacturing Co., a mill in western Maine. "Wool fibers will compost, but not easily," he says. "We can get the heat of the compost pile up to nearly 55° C, but not quite all the way yet. Mostly the problem is getting the moisture down." Brinton gives two good reasons why woolen mills are looking closely at composting. "First, it is difficult in Maine to get a land application permit. If your material is composted, part of the permit requirements are waived. Secondly, wool fiber is high in nitrogen. The composted material runs about seven percent nitrogen."

One Million Tons

The head of the Massachusetts Food and Agriculture Department maintains that "one million tons of the state's solid waste total of 6 million tons can be composted economically. Materials like yard wastes, apple pomace, fish processing wastes, racetrack manure, waste paper from paper companies and a lot more." That's why Gus Schumacher has set up a series of private sector ventures with grants from his department. Five companies were in the program started in 1986.

The Massachusetts Natural Fertilizer Company in Westminster plans to market a fertilizer made from chicken and horse manure, paper sludge and wood chips. BioCom Limited, based at Appleton Farms in Ipswich, markets 1,000 tons of compost made from horse and cow manure, and crab gurry.

34

BIOLOGICAL TREATMENT OF ORGANIC TOXIC WASTES

George M. Savage, Luis F. Diaz and Clarence G. Golueke

TWO very convincing reasons can be advanced in favor of the feasibility of composting as an option for treating biodegradable hazardous wastes. The first is the impressive array of organic chemical compounds, some of which are hazardous, that has been demonstrated in laboratory studies to be susceptible to breakdown by microbes isolated from enrichment cultures. The second is an abundance of experience that has demonstrated the breakdown of many organic compounds when incorporated into the soil. The conclusion to be drawn from these two facts is that if organisms found in nature can break down a hazardous compound in the laboratory or in the soil, they should be able to accomplish the breakdown in a compost environment. The validity of this conclusion is confirmed to some extent by the successes attained in the past in the composting of certain hazardous wastes.

At this juncture, it should be pointed out that the emphasis in this paper is on wastes that have been rendered hazardous by virtue of having been contaminated by a toxic substance. The aim is to render such wastes harmless by destroying the toxic contaminants, much as composting is used to destroy pathogens in wastes contamined with them. Nevertheless, in the discussion that follows, the terms "substrate" and "organics" may be used occasionally for referring to the hazardous contaminant alone. Because pesticides and oily sludges have been demonstrated to be biologically degradable, the greater part of the sections that follow deal with those two groups of wastes.

An extensive literature survey covering about sixty studies of microorganisms able to assimilate recalcitrant organic materials is presented in a report edited by D.J. De Renzo[1]. In the course of the survey, it is pointed out that: 1) relatively little information is available in terms of descriptions of microorganisms capable of metabolizing recalcitrant materials; and 2) most of the studies dealt with monocultures and idealized carbonaceous substrates applied under carefully controlled laboratory conditions. Therefore, although information gained in those studies may perhaps not be directly applicable to the complex conditions characteristic of the compost process, it can serve as a foundation for studies of complex interactions among microbial populations, and for selecting a suitable microbial "seed" either for starting or accelerating the composting of degradable organic wastes.

Apparently, hydrocarbon substrates can be assimilated by more than 200 species of microbes. Indeed, fungi have been observed growing in jet fuels[2] and in toluene[3]. Examples selected from this assortment are listed in Table 1; and in Table 2 are presented examples of other organic compounds that can be metabolized together with the microorganisms responsible for the metabolism.

Land Treatment Experience

Land treatment of certain of the hazardous wastes has become a relatively common practice since its introduction in the U.S.A. as a treatment technique for petroleum sludges[14]. As of May 1983, from 1 to 3 percent of the hazardous wastes currently being generated was land-

TABLE 1

Microbial Groups Capable of Metabolizing a Hydrocarbon Substrate

Bacteria		Actinomycetes	Fungi
Achromobacter	Flavobacterium	Actinomyces	Aspergillus
Aerobacillus	Gaffkya	Debaryomyces	Aureobasidium
Alcaligenes	Hansenia	Endomyces	
Arthrobacter	Methanomonas	Nocardia	Cephalosporium
Bacillus	Methanobacterium	Proactinomyces	Cunninghamella
Bacterium	Micrococcus	Saccharomyces	Mycelia (Fungi imperfecti)
Beyerinckia	Micromonospora		Monila (yeast)
Botrytis	Mycobacterium		Torula (yeast)
Candida	Mycoplana		Torulopsis (yeast)
Citrobacter	Pseudomonas		Trichoderma (yeast)
Cellustomonas	Sarcina		
Colostridium	Serratia		
Corynebacterium	Apicaria		
Desulfvibrio	Spirillium		
Enterobacter	Thiobacillus		
Escherichia	Vibrio		

treated at the approximately 200 sites in operation[15]. Type of user and extent of use for particular wastes are indicated by the list in Table 3. Judging from the data in the table, land treatment of petroleum refining wastes far surpasses that of other uses.

Compost Experience

The number of reports in the literature on the composting of hazardous wastes is very small. Probably the first such report is that by Rose and Mercer[16], who investigated the composting of insecticides in agricultural wastes. They found that the concentrations of diazinon and parathion are rapidly decreased by continuous thermophilic composting. Thus, within 10 days the concentration of diazinon was only 50 percent of its original concentration, and within 42 days it was less than 0.002 ppm as compared to its original 3.3 ppm. The degradation of parathion amounted to 50 percent of its original concentration in 12 days; and that of pp-DDT had reached about 65 percent at the end of 50 days. The DDT breakdown products DDD and DDE could not be found.

In an investigation of the compostability of a 5 to 10 percent TNT mixture, Osmon and Andrews[17] found that the TNT could be reduced to zero or an acceptable level within a relatively short time. In a project involving the composting of sewage sludge, Energy Resources Co. Inc.[17] found that many of the aromatic compounds in the sludge were reduced to negligible levels. Preliminary results of research done by Cal Recovery Systems, Inc. indicate that concentrations of Sevin (carbaryl) in a composting mass are rapidly reduced to zero; and the oil in oil sludge, to less than 75 percent of its original concentration in less than a month. In keeping with land-farming

TABLE 2

Organic Compounds that Can be Metabolized, and Responsible Microorganisms

Compound	Microorganism	Reference
Phenylmercuric acetate	Pseudomonas, Athrobacter Citrobacter, Vibrio	4
Raw rubber, hevea latex	Actinomycetes, bacteria	5
Detergents	Nocardia, Pseudomonas	6
PCB's	Not identified	7
Malathion	Trichoderma, Pseudomonas	8
Endrin	Arthrobacter, Bacillus	9
Lindane	Clostridium	10
DDT	Escherichia, Hydrogenomonas Saccharomyces	11,12
Dieldrin	Mucor	13

TABLE 3

Major Users of Land Treatment

User	No. of Facilities
Petroleum refining	100
Industrial organic chemicals	9
Wood preserving	6
Petroleum production	4
Plastics, materials, residues	3
Paints and allied products	3

experience in general, the breakdown of the oil occurred only when the environment was aerobic.

"Seeding", Substrates and Mechanisms of Breakdown

Composting hazardous wastes differs from routine composting in that "seeding" of the raw wastes is an important feature. The "seeding" is done by recycling a portion or even most of the compost product into the raw waste to be composted. A suitable ratio would be on the order of one part "seed" to nine parts raw waste. The seeding ensures the presence of a population of organisms capable of attacking the hazardous contaminant without incurring the need for a lag period during which the necessary population could develop. In effect, an enrichment system is established.

A perusal of the literature noted previously reveals that almost all substrates organic in molecular structure (i.e., "organic" in the broad chemical sense) are amenable to biological breakdown. The implication is that biological origin is not a requisite for biodegradability. As stated earlier, given the right conditions, hydrocarbons are broken down relatively readily. Aromatics are degraded more slowly than aliphatics and alicyclic nonaromatics. Microbes have enzymes (oxygenases) that convert polynuclear aromatics into unstable cyclic peroxide, which in turn is reduced to *cis*-glycol. The *cis*-glycol is further oxidized. It should be noted that the researchers who determined this pathway, contend that in land treatment, a period of 1 to 2 years might more closely resemble the potential of soil to decompose these compounds.[18]

Not all of the breakdown in soil and hence in a compost pile is biological in nature. Chemical and physical conditions established by the compost process can exert an important influence on the breakdown of certain organic compounds. Ultraviolet light, temperature, and pH level are three such conditions. Many biodegradable pesticides are so vulnerable to these factors, that isolating and determining the role of microorganisms in the breakdown of the pesticide is very difficult[8,19].

Temperature and pH levels reached during the composting process are those at which such pesticides decompose most rapidly.[19] Hence, in composting, physical, chemical, and biological destructive agents combine to degrade the pesticides, thus magnifying the effectiveness of composting.

Advantages of Composting

An evaluation of the advantages of composting should be prefaced by calling attention to a very important proviso, namely, that advantages offered in favor of composting particular hazardous compounds are applicable only if the compounds are susceptible to biological breakdown. One approach to the evaluation is to judge in terms of the superiority of composting to incineration and landfarming. Accordingly, the evaluation that follows, begins with comparison of composting with the two methods. It begins by comparing the advantages to be gained by composting toxic organic compounds with those from incineration or landfarming.

Toxic Organics

Because the volumes usually involved are large, composting contaminated residues is less expensive than incineration in terms of operation and equipment. Moreover the threat to the environment is less severe. Furthermore, composting can be carried on at generation sites with less effort than is required with portable incineration units.

In comparison with simple storage and land treatment (landfarming), the composting approach is safer and is competitively inexpensive. The competitive advantage is gained by accelerating the rate of biological destruction, which occurs much more slowly in nature, and by generating destructive levels of key physical and chemical factors. For example, less persistent pesticides such as malathion and carbaryl break down rapidly at pH levels and temperatures routinely attained in composting.

With respect to halogenated organic compounds, the anaerobic phase needed to initiate breakdown is readily attained by failing to aerate the composting mass at some stage of the process—preferably the first stage. Composting is safer than landfarming because the material is contained during the compost process rather than being spread or worked into the land. The containment makes it possible to control and treat all emissions before releasing them to the environment.

A final advantage, albeit not always attained, is the conversion of the waste into a useful product, namely the compost product. However, for the product to be safely used, the toxic contaminant and toxic breakdown products must have been completely destroyed or at least must have been reduced to an acceptable level.

Oily Wastes

With respect to oily wastes, composting most likely fares better when the comparison is with landfarming than with incineration. Nevertheless, the usual argument of lower capital and operating costs can be made in favor of composting in preference to incineration. However, because of the scarcity of firm data on the costs of composting oil wastes, the comparison herein is confined to composting versus landfarming. Another reason for concentrating on landfarming is that it is fairly commonly used for the treatment of sludges generated in petroleum refining.

As stated earlier, although landfarming has been proven to be quite effective in the destruction of oily substances, the size of the required land areas is dauntingly large. This is where composting comes in. Through composting it is possible to enjoy the advantages of biological mechanisms at a minimum dedication of land.

It is true that a "steady state" can be reached in landfarming by providing an area sufficiently large to permit the waste to be applied in successive daily or weekly increments, such that by the time the land area has been completely covered, the oily wastes in the first increment have been satisfactorily broken down. In which case, the succession can be continuously repeated and a "steady state" is reached. The drawback is the size of the field required for reaching a "steady state". Because the time element is much shorter with composting, the land area required for the composting operation is far smaller. The shortening of the time requirement is due to the acceleration of the biological breakdown processes brought about in composting.

Land area is but one of the many advantages of composting over land-farming. An important advantage is the complete control of all emissions that can be attained in composting as contrasted to the rather relaxed characteristic of landfarming. Unless expensive measures are taken in landfarming to minimize the generation of leachate and to collect that which is generated, the possibility of contaminating ground water becomes serious. Equally serious is the potential for polluting the air with hydrocarbon emissions. In fact, the EPA regards this aspect of landfarming as reason for serious concern.

Yet another important advantage of composting is the sharp reduction in amount of residue in need of final disposal. The amount of this residue can be made quite small by using the compost product as a bulking agent, and as such recirculate it within the system. Then, too, there is the possibility that the final residue can be used under careful control in landscaping around the facility.

Technology

Through suitable modifications, existing technology can be applied to the composting of hazardous biodegradable wastes. However, when choosing among the available systems, performance of the candidate systems with respect to emission control is a major consideration. Regarding emissions, the degree of importance to be attached to gaseous emissions is a principal difference between composting hazardous wastes and that of conventional wastes ("routine composting"). Whereas in "routine composting" the emission of objectionable odors is the chief concern, in composting hazardous wastes, odor ranks second to the emission of volatile toxic compounds. The relative seriousness of the preceding two types of gaseous emissions needs no explanation. The point is that in "routine composting" malodors constitute a nuisance, whereas in hazardous waste composting, volatilized toxic compounds pose a hazard not only to workers but also to the public at large. The matter of volatile substances is especially serious when composting petroleum industrial wastes.

Obviously, the urgent need to control the gaseous emissions from composting hazardous wastes demands the use of a compost system especially designed to meet it. Therefore, one of the modifications referred to in the opening sentence of the preceding paragraph would be to install a mechanism for collecting and treating the gaseous emission. The type of mechanism for the task would depend upon the nature of the gaseous substance or substances. It could be based on absorption or adsorption, on combustion, or on a chemical reaction.

An important constraint is the necessity of confining objectionable gaseous emissions within the reactor itself until the emissions are processed, isolated, or removed. This constraint rules out arrangements in which the only provision for gaseous emission control is by way of the ventilation system of the building that houses the reactor. The reason is readily apparent: Although the gaseous emissions may not reach the environment external to the building, they do affect individuals (i.e. workers) within the structure. This constraint also militates against the use of the conventional forced aeration (static pile) system used for sewage sludge treatment. Most of the current compost reactors (drum, silo, tank) probably could be suitably modified to fit these more stringent requirements.

Practically all existing compost systems are readily adaptable to caring for the remaining two classes of emissions, viz., aqueous and solid. The corollary is that if there are no toxic gaseous emissions with which to contend, almost all existing systems would be suitable, provided, of course, that due safeguards be instituted with respect to handling and dust generation.

Unless halogenated organic compounds are to be destroyed, an aerobic system would be the one of choice. If the halogenated compounds are to be composted, then an anaerobic phase would have to be included.

Feedstock preparation would depend upon the nature of the waste as a whole (waste + toxic contaminant), and would be the same as that for conventional composting. If an oily sludge is to be composted, a bulking agent would be required. Bulking agents used in sewage sludge composting would be satisfactory.

Prognostications

When conjecturing on the future of hazardous wastes composting, attention must be given to the advantages and disadvantages of composting as compared to those of physical, chemical, and thermal methods. The economic advantage responsible for the spread of sewage sludge composting also holds true with the composting of hazardous agricultural residues and petroleum sludges. The economic advantage over incineration is a result of the sizeable volumes of the agricultural residues involved, and because of the difficulties and expenses attending the incineration of petroleum sludges. With respect to the latter, an extensive record of successful landfarming has paved the way for a switch to composting. The switch is being spurred by the increasing stringency of governmental regulations. As long as these regulations can be more cheaply met through composting than through incineration, composting will gain in favor with the petroleum industry.

The full potential of composting in the treatment of hazardous wastes can be realized only after a thorough and intensive and extensive program of research and study has been completed. The reasons are that past experience has been sketchy and much remains to be learned.

References

1. *Biodegradation Techniques for Industrial Organic Wastes,* editor, D.J. De Renzo, Noyes Data Corp, Park Ridge, N.J., 1980 (pp. 10-103).
2. Prince, A.E., "Microbiological Sludge in Jet Aircraft Fuel", *Dev. Ind. Microbiology, 2*:197 (1961).
3. Nyns, E.J., J.P. Auguiere, and A.L. Wiaux, "Taxonomic Value of the Property of Fungi to Assimilate Hydrocarbons", *Antonie van Leeuwenhoek J. Microbiol. Serol., 34*:441-457 (1968).
4. Nelson, J.D., W. Blair, F.E. Bringkman, R.R. Colwell, and W.P. Iverson, "Biodegradation of Phenylmercuric Acetate by Mercury-Resistant Bacteria", *Applied Microbiology, 26*:321-326 (1973).
5. ZoBell, C.E. "Assimilation of Hydrocarbons by Microorganisms", *Adv. Enzymol., 10*:443-486 (1950).
6. Mitchell, R., *Introduction to Environmental Microbiology,* Prentice-Hall, Englewood Cliffs, N.J., 1974 (418 pp).
7. "Bugs eat their way through PCB's", *Chemical Engineering, 92*(3):21 (Feb. 4, 1985).
8. Paris, D.F., D.L. Lewis, and N. Lee Wolfe, "Rates of Degradation of Malathion by Bacteria Isolated from Aquatic Systems", *Environ. Science and Technology, 9*(2):135-138 (Feb. 1975).
9. Patil, K.C., F. Matsumura, and G.M. Boush, "Degradation of Endrin, Dieldrin, Aldrin, and DDT by Soil Microorganisms", *Appl. Microb., 19*:879-881 (1970).
10. McRae, J.C., K. Ragdu, and E.M. Bautista, "Anaerobic Degradation of the Insecticide Lindane by *Clostridium* sp.", *Nature, 221*:859-860 (1969).
11. Focht, D.D., "Microbial Degradation of DDT Metabolites to Carbon Dioxide, Water, and Chloride", *Bull. Environ. Contam. Toxicol., 7*:52-56 (1972).
12. Kallerman, B.J. and A.K. Andrews, "Reductive Dechlorination of DDT to DDD by Yeast", *Science, 141*:1050-1051 (1963).
13. Anderson, J.P., E.P. Lichtenstein, and W.F. Whittingham, "Effect of *Mucor alternans* on the Persistence of DDT and Dieldrin in Culture and Soil", *J. Econ. Entomol., 63*:1595-1599 (1970).
14. Grabowski, T.M. and A.J. Raymond, "Disposal and Transportation of Refinery Wastes", in *Hazardous and Toxic Wastes: Technology, Management and Health Effects,* Pennsylvania Academy of Sciences, 1984.
15. Morrison, A., "Land Treatment of Hazardous Waste", *Civil Engineering Magazine,* pp. 33-38 (May 1983).
16. Rose, W.W. and W.A. Mercer, *Fate of Insecticides in Composted Agricultural Wastes,* Progress Report, Part I., Nat'l. Canners' Association, Washington, D.C., 1968.
17. Epstein, E. and J.E. Alpert, "Composting of Industrial Wastes", in *Toxic and Hazardous Waste Disposal,* Vol. 4, Ann Arbor Science Publishers, Michigan, 1980.
18. Overcash, M.R. and D. Pal, *Design of Land Treatment Systems for Industrial Wastes—Theory and Practice,* Ann Arbor Science Publishers, Inc., 1979 (pp. 246-255).
19. Mount, M.E. and F.W. Oehme, "Carbaryl: A Literature Review", *Residue Reviews, 80*:1-64 (1981).

35

COMPOSTING FOOD INDUSTRY WASTES

David Riggle

NEW DEVELOPMENTS in the state of Washington point to both the future and the past. The town of Grandview and neighboring Prosser are presently siting a new landfill in arid southern Washington. Since a number of food processing industries are located in the area, approximately 30 percent of the two towns' combined waste stream consists of materials like apple pomace, grape skins and cranberry residuals. Officials in Grandview and Prosser applied to the Washington Department of Ecology for a recycling grant to develop a facility for composting the food waste along with yard debris at their new landfill. Based on the 3,714 tons of food processing waste out of the total 12,380 average tons coming to the landfill every year, the towns (population about 6,000 each) estimate they can save about $10,000/yr in diverted landfill costs.

The Department of Ecology reacted favorably to their request, but decided to grant them $26,250 of an estimated $35,000 necessary to do a Food Waste Composting Feasibility Study before funding 75% of the $100,000+ needed to actually implement the composting project.

According to Peter Christiansen of the Washington DOE, his department would very much like this project to be a model for the rest of the state. If Grandview and Prosser agree to do the feasibility study, and the results point favorably toward composting the industrial food waste, the state will have set an example that other communities can follow if they choose.

The proposed project, as outlined in the draft sent to Grandview and Prosser by the Washington DOE in March 1989, "would consist of an economic and technical feasibility study to determine if the organic waste stream currently generated in Grandview/Prosser would be economically processed [by composting] to achieve maximum materials recovery. The study will also attempt to address technical issues such as vectors, odor problems, and existing technologies for materials handling and recovery."

The major project tasks for the study are outlined as follows:

1. Determine the full composition and sources of waste stream including seasonal fluctuations, and identify and quantify potentially compostable food waste, such as apple pomace, grape pulp, yard and garden waste, and other organic materials.
2. Perform a technical and economic analysis for the moisture content, total nitrogen and carbon of each waste type; identify residues of pesticides used by growers; determine availability of bulking agents; analyze capital costs, operational and maintenance costs, potential revenues and avoided costs; examine uses and long-term markets for the compost product; and project tipping fees.
3. Determine the optimum technologies for the facility envisioned and identify alternative systems, with particular attention paid to potential problems and solutions with odors, carbon/nitrogen ratios, moisture, vectors and contaminants.
4. Develop a conceptual design for the composting system including size and type of machinery, access require-

ments, structural design and layout, storage, safety and utilities.

5. Survey food industries, farmers and targeted residents in the jurisdiction and determine commitment and interest in the project; and prepare a report summarizing all of the above by June 15, 1990.

At the time of this writing in mid-April, 1989, the cities of Grandview and Prosser are deciding if they will accept or modify the Washington DOE's suggested proposal.

Research Into Composting Food Processing Wastes

Perhaps the most extensive research conducted on the technical aspects of this process was by the National Canners Association in Berkeley, California during the 1960's. That Association changed its name to the National Food Processors Association in 1977. Walter W. Rose, now Associate Director of Research for Environmental Engineering, was on the team that did the research into food waste composting over 20 years ago.

"At the time of the study, processors in the East Bay area were facing problems. Landfills were getting full and they were looking for alternatives. Our research indicated that composting offered a feasible and esthetically acceptable method for disposal of high-moisture wastes, such as fruit and vegetable solids, but we didn't really have a good feel for the economics back then, or what to do with the end-product. Today there is a market for compost, but when we did the research there was none."

Detailed results of the National Canners Association studies appeared in *Compost Science* magazine (the original name of *BioCycle*) in Autumn, 1962 ("Aerobic Composting of Vegetable and Fruit Wastes," pp. 9-19), Summer, 1965 ("Composting Fruit and Vegetable Refuse," pp. 13-25), and Autumn, 1968 ("Windrow Composting of Fruit Waste Solids," pp. 19-21). Additionally, a report by J.F. Kefford of Australia's Commonwealth Scientific and Industrial Research Organization appeared in Autumn-Winter, 1966 ("Composting Trials with Pear Canning Wastes," pp. 24-26). Dr. Kefford visited the research trials in Berkeley and duplicated their methods and essential results after returning to Australia.

Summary of Experiments

Following is a summary of the composting experiments performed by the National Canners Association:

Bin Type Experiments—Fruit Waste

1. Dry materials such as municipal composts or rice hulls were used to absorb the moisture from fruit waste solids and adjust the moisture concentration of the mixture within the range optimum for high rate composting.

2. White pine sawdust and coffee ground residue from the manufacture of instant coffee performed satisfactorily in absorption of moisture and maintenance of porosity of the compost pile; however, the strong buffering system in each of the materials depressed changes in pH and prolonged acidic conditions within the mixture.

3. The strong buffer system within the sawdust and coffee grounds was destroyed after one complete compost cycle, making the finished material more suitable as an absorbent in additional experiments.

4. Of the other dry admixtures used, shredded redwood bark and wheat straw failed to provide conditions necessary for the composting of fruit wastes. Both substances failed to adequately absorb moisture and maintain compactness to a degree required to prevent heat losses from the pile.

5. For fruit waste-dry material mixtures, the optimum initial moisture concentration was between 60 and 70 percent.

6. The maximum weight ratio of fruit to dry material successfully composted was 250 pounds of peach solids to 100 pounds of municipal compost. More efficient composting conditions were obtained when the weight of the fruit was reduced to 200 pounds.

7. When peach solids were ground or disintegrated prior to composting, the weight ratio of fruit to dry material had to be reduced, since grinding released liquid from the tissues and gave a higher initial moisture concentration for the mixture. However, under optimum conditions, grinding of the fruit reduced by approximately 50 percent the time required for stabilization of unground fruit.

8. Aerobic conditions in the composting mixture were maintained by a turning schedule in which an initial turning was done 24 hours after layering the fruit and dry material into the bins, followed by daily turning for 5 to 6 days and turning thereafter on alternate days until the process was completed.

9. The initial low pH (4.5 to 5.0) of fruit-compost mixtures caused a lag in the development of desirable microbial microflora and elevation of temperature. Lime was added to neutralize some of the acidic components. Following the addition of lime, the pH rose toward neutral, with a resultant increase in temperature values.

10. The addition of nitrogen materials to the compost mixture stimulated microbial growth resulting in shorter compost cycles.

11. Because fruit waste solids were approximately 85 percent moisture, evaporation during the composting and microbial conversion of carbohydrates and acids to carbon dioxide and water, significant losses in weight and volume occurred. Reductions ranged up to 70 and 59 percent, respectively, for the weight and volume of the initial mixtures.

12. The finished, air dried compost material was recycled for use as the absorbent material for fresh waste additions.

13. Offensive odors did not develop during the composting of fruit waste solids. Slight fruit fermentation odors were noticeable at the beginning of the compost cycle followed by ammonia odors during the later stages of composting.

14. Fly breeding in the compost was not a problem. Fruit pieces and possibly the warmth of the surface layer attracted flies but did not result in egg deposition.

Bin Type Experiments—Produce-house Waste

1. Produce-house waste such as that discarded by grocery stores can be composted when mixed with a dry material such as rice hulls.

2. When the waste is added frequently to an actively decomposing mixture, inhibitory or toxic substances gradually accumulate in the mixture.

3. After 70 days of operation the experiment was terminated because there was no further degradation of the waste.

4. Water soluble, heat resistant materials were extracted from the finished material which inhibited growth of bacterial vegetative cells and germination of spores.

Open Windrows

1. Windrow composting was found to be a feasible method for study of the disposal of large quantities of fruit waste.

2. Fruit waste was added at frequent intervals to an active compost mixture without any observed detrimental effects.

3. Height of the rice hull windrow was important in preventing heat losses and maintaining high temperatures.

4. Open windrows of rice hulls could not be maintained beyond 4 feet in height. Above this height, the base of the windrow extended beyond the reach of the turning mechanism.

5. Windrows could be turned by machine for the purpose of mixing and aerating the compost mass.

Closed-wall Windrows

1. With stationary walls, the windrow height of rice hulls could be extended to 5 to 6 feet, resulting in higher internal temperatures.

2. Continuous composting at thermophilic temperatures was shown to be a feasible method for the disposal of fruit wastes.

3. With high-temperature composting, air-injection into the compost mixture was necessary to maintain aerobic conditions.

4. Daily turning, without forced aeration, did not provide sufficient aeration to prevent the formation of anaerobic pockets within the windrow.

5. The production of heat in the windrow after turning was sufficiently rapid to permit adding of fresh waste and turning on a 12-hour interval, resulting in the disposal of more waste than turning and adding on a 24-hour time cycle.

6. A temperature gradient developed in the windrow with the highest temperatures located near the surface of the windrow and the lowest temperature located near the bottom.

7. Heating of air prior to injection into the windrow decreased the range or spread in the temperatures taken at various depths.

8. Lime was added at the beginning to increase the pH of the windrow mixture after too much waste had been added, causing the temperature and microbial activity to increase.

9. An automated system was developed to handle the waste, grind and transport the material to the windrow, to add the waste, and to turn the windrow.

Municipal Perspective for the '90s

The cities of Grandview and Prosser are approaching composting food processing wastes from a municipal rather than an industrial perspective, but, according to their initial proposal, local industries have made numerous inquiries regarding composting of food processing wastes and there is a genuine commitment to this program.

Mr. Rose wasn't aware of anyone in the food processing industry actively composting at the present time. In fact, he knew of very little that had actually happened as a result of the National Canners Research. "Eventually a new landfill was found in the Bay Area and that was the end of that," he says.

The National Food Processors Association's publication, "1988 Report of NFPA Technical Activities," includes an entry on Municipal Solid Waste, but makes no mention of composting whatsoever. It states, ". . . now is when society must consider the options for meeting our future solid waste management needs—recycling, waste to energy conversion, landfilling and source reduction." This section of their report is mainly focused on the "packaging challenges" faced by the food processing industry. "In particular the industry is interested in assuring that national guidelines will govern the use of packaging materials so as to avoid disruptions to interstate commerce that would accompany regulations at the state or local level," but the report does mention that an ad hoc Municipal Solid Waste Issues Committee has been formed to develop an industry policy on MSW. Perhaps they will examine composting as a valid disposal option if municipal projects like the one proposed in Washington state are successful.

Mr. Rose notes that the research of 20 years ago really only needs to be demonstrated and refined using current technology to show the economics and end uses more clearly.

SECTION NINE

THEORY, PRACTICE AND SUMMARY

36

OPERATING AND COST DATA FOR IN-VESSEL COMPOSTING

John Johnston, John F. Donovan and Albert B. Pincince

DURING THE LAST 15 years, composting has grown in popularity as a sludge management technique, in part due to higher costs and greater environmental concerns often associated with alternative methods. Compared to static pile and windrow composting operations, in-vessel systems promise a more stabilized and consistent product, smaller space requirements, and better containment and control of odors. Because in-vessel composting systems are new in this country, however, there is relatively little information available on the operation and costs of actual systems.

With funding by the U.S. Environmental Protection Agency, Camp Dresser and McKee Inc. is conducting a technology evaluation of in-vessel composting. The purposes of this project are to collect, evaluate, and disseminate information on full-scale in-vessel composting systems and to provide guidance on the parameters important in the procurement, selection, design, and operation of a successful in-vessel composting project.

This report presents some of the operating and cost data that were gathered during site visits to eight operating facilities in the period from April through November of 1988. Odors and materials-handling were posing problems for these systems, but all of the plants surveyed had been shown to be capable of making usable compost products.

System Status

Basic information on the eight plants visited is summarized in Table 1. All six in-vessel equipment suppliers with full-scale installations in the U.S. were represented. For two suppliers—Taulman-Weiss and Purac—examples of both large and small plants were included.

In four of the eight plants, both "composting" and "curing" (i.e., first-stage and second-stage composting) were done in reactors. At two plants, in-vessel composting was followed by aerated static pile curing. No formal curing step was employed at two plants. At most plants, the composted material was stored unaerated, for some additional minimum period of time before release to the public. The minimum storage times were generally established by state regulatory agencies.

The status of each composting plant at the time of the site visit is shown in Table 2. The completion of acceptance testing and start of "commercial" operations occurred from 5 to 11 months after the start-up dates shown in the table.

Since start-up, many of the plants have been shut down for modifications or repairs. The odor control and aeration systems have been modified at Cape May. The Schenectady plant was taken out of service for structural repairs and upgrading of the odor control system. Operations at Newberg have been interrupted twice for structural repairs. The Portland plant was taken out of service for a month in 1988 so that miscellaneous modifications could be accomplished and so that the waste activated sludge storage lagoon could be dredged.

Only three plants—Clayton County, Plattsburgh, and Sarasota—were processing all of the sludge available or

TABLE 1.
Plant Data

Location	Type of Reactor	Supplier	System Configuration First Stage	Second Stage
Akron, OH	Rectangular agitated bin	Compost Systems Co. (Paygro)	In-vessel	None (aerated curing under study)
Cape May County, NJ	Rectangular vertical reactor	Purac Engineering	In-vessel	Aerated static pile[1]
Clayton County, GA	Cylindrical vertical reactor	Taulman-Weiss	In-vessel	In-vessel
Newberg, OR	Rectangular horizontal tunnel	Ashbrook-Simon-Hartley	In-vessel	Aerated static pile
Plattsburgh, NY	Circular agitated bin	Compost Systems Co. (Fairfield)	In-vessel	None
Portland, OR	Cylindrical vertical reactor	Taulman-Weiss	In-vessel	In-vessel
Sarasota, FL	Rectangular vertical reactor	Purac Engineering	In-vessel	In-vessel
Schenectady, NY	Rectangular vertical reactor	American Bio Tech	In-vessel	In-vessel

Notes:
[1]System constructed with in-vessel curing but changed to configuration shown in 1987-88.

intended for each plant. The processing rates of the other five facilities were limited for one reason or another. The Akron plant was operating at reduced capacity to ameliorate local odors while new odor control equipment was being piloted. At Cape May, the new odor control system was in place, but return to full plant capacity was being done in stages under the direction of regulatory authorities. The Portland plant has been hampered by mechanical problems of various kinds that have prevented simultaneous operation of all six reactors for extended lengths of time. The Newberg plant was conducting acceptance testing at the time of the site visit. It has since been shut down because of a structural failure. The acceptance testing at Schenectady has been completed, but the plant was out of service for correction of structural problems and upgrading of the odor control system.

Design capacities and actual sludge processing rates are also shown in Table 2. At Cape May, the cure reactor was converted into a first-stage composting reactor when aerated static pile curing was initiated. This modification increased the plant capacity from 12 dry tpd as designed to 20 dry tpd.

Performance

All of the plants were easily meeting the EPA time-temperature requirement for a pathogen-reducing process (55 C for 3 days) in the first stage reactor. Other performance data for the eight plants are shown in Table 3.

All of the plants use sawdust as amendment. Akron uses shredded bark as well. With the exception of Sarasota, all of the plants are using a larger fraction of amendment in their mix ratios than was anticipated in design. Reasons cited for increasing the amendment were to improve the handling characteristics of the mix, to increase the porosity of the mix, and/or to compensate for mix components with higher moisture contents than expected. In practice, the creation of an acceptable mix has been an empirical process that depends on the particle size distributions and textures of the amendment and recycle materials, as well as their moisture contents. Sawdust costs ranged from $3 to $14.25 per cubic yard.

The solids content of the composted material increased by less than 5 to over 20 percentage points in the first-stage reactor. The increase in solids content was a function of both the aeration rate per unit volume of reactor and the "plug flow" detention time of the material in the reactor. The solids contents of compost products (before unaerated storage) ranged from 40 to 71 percent.

Marketing success varied from plant to plant. Every plant experienced a period of product accumulation at the beginning of operations. The reasons for the initial accumulation of compost included delays in getting state permits for product distribution, unwillingness on the part of brokers or users to contract for material of varying quality

TABLE 2.
System Status at Time of Site Visit

Location	Start-up Date	System Status	Sludge Processing Rate (dry tpd) Design	Actual
Akron, OH	Dec. 86	Operating at reduced capacity because of current odor problems	73	25[1]
Cape May County, NJ	May 85[2]	Operating at reduced capacity because of past odor problems	20 (12.1)[3]	11 11
Clayton County, GA	Aug. 86	Processing all the sludge available	2.9	1.4
Newberg, OR	Aug. 87	Undergoing acceptance testing	3.5	—
Plattsburgh, NY	Mar. 86	Processing all the sludge available	34	28
Portland, OR	Mar. 85	Operating at reduced capacity because of mechanical problems	60	26[4]
Sarasota, FL	Aug. 87	Processing all the sludge available	6.3	4.3[5]
Schenectady, NY	July 87	Shut down for modifications to structure and odor control systems	15	—

Notes:
[1]Seven day/week basis. Actual loading is 44 dtpd, 4 d/wk.
[2]Start-up date for original configuration.
[3]Plant modifications made in 1987-88 increased the capacity to 20 dry tpd by adding aerated static pile curing and converting the existing cure reactor to a first-stage composting reactor.
[4]Average for first six months of 1988.
[5]Annual average, 3.9 dtpd in summer; 4.7 dtpd in winter.

TABLE 3.
Performance Data

Location	Volumetric Mix Ratio Sludge/Amendment/Recycle	Sludge	Solids Content Initial Mix	(% by weight) First Stage	Second Stage
Akron, OH	1/0.4/0.6/2[1]	20-26	40-50	60-68	—
Cape May County, NJ	1/0.6/1	20-30 (avg: 23)	37	53-65	—
Clayton County, GA	1/1.4/1.4	16-18	34-35	39-44	45-52
Newberg, OR[2]	1/0.5/7.5	16-18	33-40	33-44	—
Plattsburgh, NY	1/1.4/3.3	20-25	38-42	43-53	—
Portland, OR	1/1.6/0.6	23-28 (avg: 25)	35-40	38-40	40-45
Sarasota, FL	1/0.6/2.4	14	34-46	50-55	60-71
Schenectady, NY	—	20-28 (avg: 25)	35-39	—	—

Notes:
[1]Sludge / bark / sawdust / recycle
[2]System was undergoing acceptance testing at time of site visit. Data shown were collected later in the test (Sept-Oct 88) after the site visit.

and production rates, and/or low initial market demand. How each facility dealt with this problem depended mainly on the amount of room available at the site. Sarasota, for instance, with essentially no storage space at the plant site, paid to have the compost hauled away and applied to land. On the other hand, Portland had sufficient space available onsite to store an inventory equivalent to two years of production.

At three of the plants—Clayton County, Portland and Akron—compost marketing was the responsibility of a subsidiary of the equipment supplier. The remaining plants were seeking or considering contracts with brokers.

The two largest user groups were landscape contractors and government agencies, who used the compost at parks, schools, highways, and landfills (as cover material). Golf courses and homeowners were taking significant quantities of compost from some plants. Local nursery industries have generally expressed interest in the product, but typically have purchased only small quantities to date for testing purposes. Compost prices varied from $2 to $19 per cubic yard, depending on local pricing policies and the quantities of compost purchased.

Problem Areas

The two major problem areas experienced by most or all of the plants surveyed were odor control and materials handling.

Of the eight facilities visited, five had records of serious odor complaints. A sixth plant had received minor odor complaints. All six plants with odor problems have operational odor control systems. The types of systems employed included wet scrubbers (one- and two-stage), ozone contact, a compost filter, and bubbling the odorous gases through water. Generation of odors did not appear to be related to the specific composting technology used.

There were several common factors that contributed to the odor problems at these plants. In at least three cases, site characteristics such as local topography, weather conditions, and the proximity of neighbors exacerbated odor problems. At some facilities, the odor control technology installed was ineffective at removing the odors or was undersized. Uncontrolled odor sources such as storage piles and open doors in the process building contributed to odor releases at several plants. The characteristics of the sludge being composted may also be an important factor. At the two plants that did not receive complaints (Portland and Newberg), the sludge being composted was relatively stable compared to the sludges at the other six facilities. The sludge at Portland was anaerobically digested prior to composting. At Newberg, the sludge was taken from an underloaded oxidation ditch with a long solids detention time.

At the time of the site visits, one plant had modified its odor control system extensively and four other plants were planning modifications. Two-stage wet chemical scrubbing was the technology most commonly being considered.

Problems with materials-handling systems were experienced at nearly every plant. Some commonly encountered problems are described below. Conveying systems used at the various plants include flat belts, cleated belts, drag chains, and screws (with and without center shafts).

Spillage, particularly at transfer points, was a common problem where belt conveyors were used. Incidental spillage was usually controlled by adjusting belt speeds, providing or modifying scrapers and cleaners, and/or by installing collection pans. At Akron, where sludge adhering to the belt material was the cause of spillage, dusting the belt with sawdust prior to loading it with sludge greatly reduced the problem.

Drag chains at several plants experienced jamming problems caused by flights that were too large or by inadequate clearance between the flights and the drag chain housing. Reducing the cross-sectional areas of the flights usually alleviated this problem.

Plugging of chutes and funnels was also a common problem. Solutions to these problems were usually site-specific. Plant operators universally recommended monitoring these locations with motion detectors tied into the central computer system controlling the materials-handling system.

Outfeed devices have presented problems at several plants with vertical reactors. These plants use screw devices adapted from other applications such as sawdust silos. The most common problem reported was excessive wear on the screw whenever screw flights were not hard-surfaced. Schenectady and Portland have both experienced drive mechanism problems, although the nature of these problems and the designs of the screws are different.

Costs

Operating costs and manpower needs for the eight plants are shown in Table 4.

Operating schedules and the number of operators required at the various plants were heavily dependent on the capabilities of the materials-handling equipment. For instance, operations staffs were smaller where the reactors could be loaded in one shift per day rather than two.

Operations personnel at most plants worked exclusively at the compost facility. Maintenance personnel at five facilities was provided by the wastewater treatment plant associated with the compost plant or by a centralized maintenance organization. Included in the "Other" category in Table 4 are supervisors, laboratory technicians, and clerical workers.

Annual operations and maintenance costs are shown in Table 4. Unless otherwise noted, the values shown reflect the 1988 budgets of the various facilities. (Because of their limited operating experiences, costs are not available for the Newberg and Schenectady plants.) Based on the sludge processing rates reported during the site visits, estimates were made of the quantity of sludge to be processed during the year. This was used to calculate the operations cost per dry ton of sludge processed. It is interesting to note that these costs exhibited an economy of scale, as illustrated in Figure 1. The plants with design capacities less than 15 dry tpd had "per ton" costs that were significantly greater than plants with design capacities greater than 15 dry tpd.

TABLE 4.

Operations and Cost Data

Location	Number of Personnel: Operations	Number of Personnel: Maintenance	Number of Personnel: Other	Operating Costs: Total Annual ($1000/yr)	Operating Costs: Per ton of sludge processed ($/dry ton, approx)
Akron, OH	8	4	2	1225[1]	135
Cape May County, NJ	9[2]	3	2	1038[3]	215[4]
Clayton County, GA	2	Maintenance by WWTP staff	1	148	290
Newberg, OR[5]	1.5[6]	Maintenance by WWTP staff	0	—	—
Plattsburgh, NY	4	3	1	1020[7]	100
Portland, OR	6[6]	3[8]	1	1100	115
Sarasota, FL	3	2.5[8]	1	600[7]	380
Schenectady, NY	3	Maintenance by WWTP staff	1	—	—

Notes:
[1]1987 expenditures.
[2]Not including 5 summer employees used for groundskeeping and utility work.
[3]1989 budget estimate.
[4]Based on expected sludge processing rate of 13.3 dtpd assuming that the plant is allowed to move to the next stage of operation. Based on actual expenditures of $474,000 (Jan.-Sept. 88) and actual sludge processing of 1200 dt (Apr.-Sept.), the unit cost was $395/dry ton.
[5]Plant undergoing acceptance testing.
[6]Operators rotate through composting plant. Value shown is an estimate of equivalent personnel.
[7]Not including sludge disposal costs.
[8]Maintenance by WWTP staff. Value shown is estimate of equivalent personnel.

FIGURE 1.

Operating Costs as a Function of Sludge Processing Rate

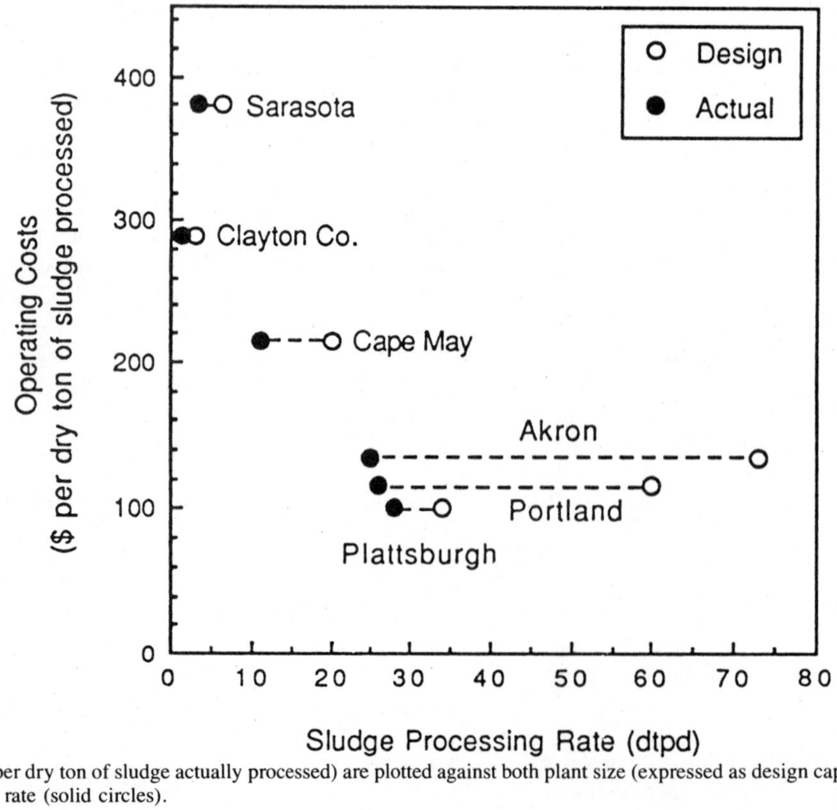

Operating costs (per dry ton of sludge actually processed) are plotted against both plant size (expressed as design capacity and shown with open circles) and actual processing rate (solid circles).

Conclusions

In-vessel composting is proving to be an effective method of producing an environmentally useful product from waste sludge. The compost being produced at all of the plants surveyed had been approved by state agencies for public use except application on food crops. While costs and operational problems have generally been greater than anticipated during design, this is to be expected in the implementation of any new, large-scale technology. As the lessons learned at these early installations are incorporated into new designs and operations, mechanical problems and costs will decline. In the future, this technology should see wide application.

Acknowledgments

The authors would like to thank the operators and other staff at the eight facilities studied. Without their generous assistance, this project would not be possible. The research described in this article has been funded by the United States Environmental Protection Agency under contract no. 68-03-3346 with Camp Dresser and McKee Inc. It has been reviewed by the Agency and has been approved for publication. Mention of trade names or commercial products does not constitute an endorsement or recommendation for use.

Authors are, respectively, senior environmental engineer, associate and vice president of Camp Dresser and McKee, based in Boston, Massachusetts.

37

CALCULATING THE CAPACITY OF A MSW COMPOSTING FACILITY

Mark Selby

WRIGHT COUNTY is located in the east-central part of Minnesota, 35 miles west of Minneapolis. Population is expected to increase from its present 73,000 to 100,000 by the year 2000. The county's solid waste management system consists of 13 refuse haulers, one demolition landfill, three sanitary landfills, one demolition disposal site, 14 recycling projects and two yard waste composting programs. In 1988, Wright County residents and businesses generated approximately 42,000 tons of solid waste or an average of 115 tons each day.

Throughout the past 4½ years, a County Task Force assessed issues, coming up with a plan in July, 1987 to initiate yard waste composting, recycling, waste reduction and public education programs. In addition, the County intends to develop a state-of-the-art solid waste composting facility which will manage the vast majority of its solid waste. The Task Force recommended that a Project Management Team (PMT) be established to guide the project's implementation. It consists of five members of the Task Force (two Commissioners and three citizens). DPRA Incorporated was selected as technical advisor to evaluate vendor systems and provide technical guidance during the design, construction, acceptance testing, and start-up of the composting facility. One of the initial technical issues that had to be addressed was the selection of a MSW processing rate or capacity for the facility.

Selection of the compost facility MSW processing rate for Wright County was difficult due to: Monthly/weekly variation[1] in solid waste generation; Projected long-term increases in amount of solid waste generated; and System economics[2] associated with a 20-year project life. This article reviews the methodology employed to assist the County in selecting a composting facility processing rate.

In determining the processing rate, factors considered included the economic impacts of operating a compost facility at various MSW processing rates and the environmental impacts of disposing of system rejects and surplus MSW. Surplus MSW was defined as monthly MSW generation that exceeds the processing rate of the compost facility during the same period.

In order to assess the environmental impacts of monthly solid waste generation variation and a forecasted increase in Wright County annual MSW generation of 32 percent over the 20 year project period, DPRA developed a computer model to estimate the amount of surplus waste and to identify periods when the facility would be under-utilized. The model incorporated monthly solid waste generation data from a study conducted for a Twin Cities Metropolitan Area County.[3] Generation data presented in this report should not be directly applied to other solid waste generation districts since this type of information tends to be region specific. The average monthly solid waste quantity and various landfill abatement activities such as yard waste composting and secondary material recycling are based upon 1990 conditions and assumed to be constant. Figures 1 through 3 graphically present the effect of selecting a facility processing at 100, 105, and 110 percent

Figure 1
MSW Generation Rates versus Compost Facility Capacity (100 percent)

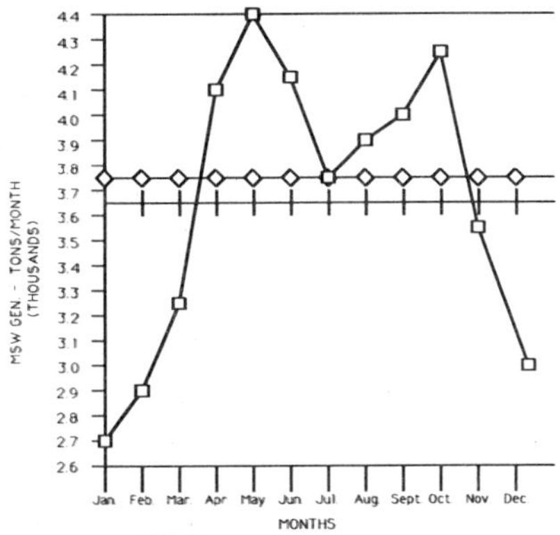

Compost Facility Capacity
170 Tons/Day at 5 Days/Week
141 Tons/Day at 6 Days/Week

Figure 2
MSW Generation Rates versus Compost Facility Capacity (105 percent)

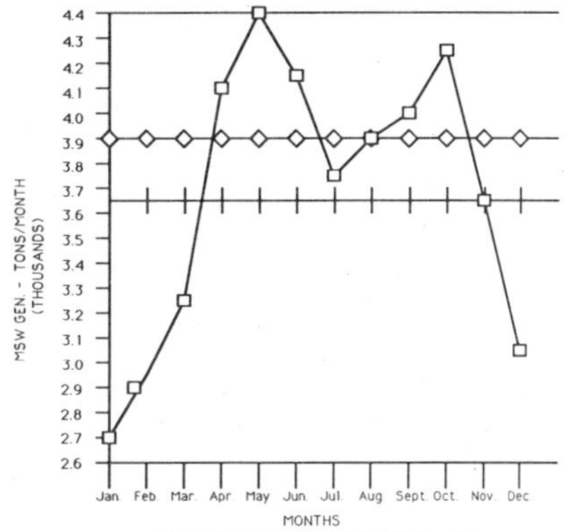

Compost Facility Capacity
178 Tons/Day at 5 Days/Week
148 Tons/Day at 6 Days/Week

Figure 3
MSW Generation Rates versus Compost Facility Capacity (110 percent)

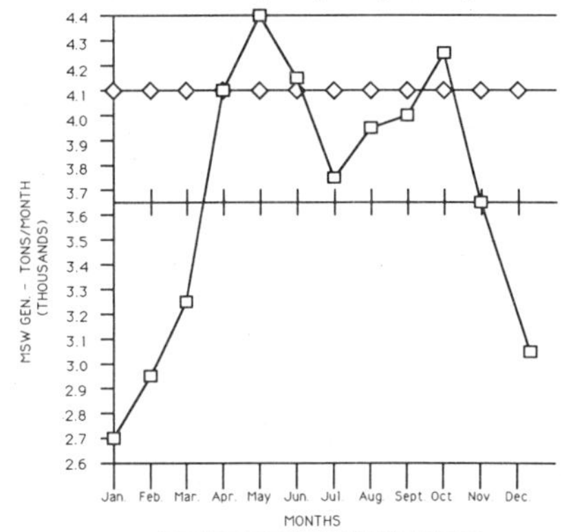

Compost Facility Capacity
186 Tons/Day at 5 Days/Week
155 Tons/Day at 6 Days/Week

LEGEND

□ ADJ. MONTHLY GEN.
\+ AVE. ADJ. GEN.
□ AVE. PLANT CAP.

PROJECTED 1990 MUNICIPAL SOLID WASTE GENERATION RATE
44,600 TONS/YEAR
3,720 TONS/MONTH

ASSUMES A CONSTANT RESIDENTIAL GENERATION RATE OF 3.2 LBS./CAPITA/DAY

ADJUSTED 1990 MUNICIPAL SOLID WASTE GENERATION RATE DUE TO RECYCLING ACTIVITIES
44,070 TONS/YEAR
3,670 TONS/MONTH

of the average 1990 monthly solid waste generation rate, respectively. Note in each case, surplus waste results and extended periods occur when the facility is under-utilized.

With the development of the surplus waste/excess plant capacity computer model, the following economic impacts can be estimated: capital investment costs; annual operating costs; tipping fees; present worth estimates; and disposal costs for system rejects and surplus solid waste streams. The following four case studies compare and analyze the environmental and economic impacts of the different processing rates.

Case I

In the first case, the solid waste processing rate is assumed to be at the early 1990s generation rate of 46,300 tons per year. It is also assumed that the facility will process 302 days per year (i.e., 6 days per week minus 10 planned holidays) and operate a single 8-hour shift. Consequently, the average resulting daily processing rate would be 153 tpd.

The capital investment and total annual costs are estimated at $7.96 million and $2.07 million, respectively. The unit processing cost for the first year and subsequent years is estimated at $44.70 per ton. The unit cost does not include the expense to transport and dispose of the process reject and surplus waste streams. The facility unit processing cost remains constant because the facility can process 46,300 tons every year even though the County's future annual waste generation rate is projected to exceed that amount.

Applying the computer model to each year from 1991 to 2010, the facility will generate approximately 278,000 tons of process rejects and 237,000 tons of surplus waste, or a total of 515,000 tons of waste during the 20 year period. The disposal cost for this waste is estimated to be $22 million (1988 dollars), assuming a disposal fee at $40 per ton for surplus waste and $45 per ton for process rejects (which includes a $40 per ton tip fee and $5 per ton handling and transportation cost).

The total solid waste management cost for the first case is $63.3 million (1988 dollars). The total system cost includes the compost facility operation and the transportation and disposal of solid waste throughout the 20 year project period. The present value over the same 20 year period is $45.5 million assuming a five percent real interest rate.

Case II

The second case uses the annual solid waste generation rate for the mid-to-late 1990s of 53,800 tons per year for the facility processing capacity. It is also assumed that the facility will process waste 302 days per year and operate a single 8-hour shift. Consequently, the resulting average daily processing rate would be 178 tons per day.

The capital investment and total annual costs are estimated at $8.80 million and $2.35 million, respectively. The estimated annual unit processing costs (not adjusted for inflation) range from $50.80, in 1991, to $43.70, after 1997 because a larger quantity of solid waste is generated and processed each year. The unit cost becomes constant (excluding inflation) after 1997 because the annual waste generation equals or exceeds the plant's processing capacity. The unit costs do not include the expense to transport and dispose of the process reject and surplus waste streams. The annual MSW generation rate is expected to exceed the facility's processing capacity in 1997 or 1998.

Applying the computer model to each year from 1991 to 2010, the facility will generate approximately 314,000 tons of process rejects and 114,000 tons of surplus waste or a total of 428,000 tons of waste during the 20 year period. The disposal cost for this waste is estimated to be $18.7 million (1988 dollars) assuming a disposal fee at $40/ton for surplus waste and $45/ton for process rejects.

The total solid waste management cost for the second case is $65.7 million (1988 dollars). The total system cost includes the compost facility operation and the transportation and disposal of solid waste throughout the 20 year project period. The present value over the same 20 year period is $48.2 million assuming a five percent real interest rate.

Case III

The third case uses the annual solid waste generation for the early-2000s assumed to be 63,800 tons per year as the facility processing capacity. It is also assumed that the facility will process waste 302 days per year and operate a single 8-hour shift. Consequently, the resulting average daily processing rate would be 211 tons per day.

The capital investment and total annual costs are estimated at $9.85 million and $2.72 million, respectively. The estimated annual unit processing costs (not adjusted for inflation) range from $58.80, in 1991 to $42.70, after 2005, per ton of solid waste received. The annual unit processing cost decreases from 1991 through 2005 because a larger quantity of solid waste is generated and processed each year. The unit cost would become constant (excluding inflation) when the plant processing capacity is equal to or less than the annual waste generation. Once again, the unit costs do not include the expense to transport and dispose of the process reject and surplus waste streams.

Applying the computer model to each year from 1991 to 2010, the facility will generate approximately 343,000 tons of process rejects and 19,000 tons of surplus waste, or a total of 362,000 tons of waste during the 20 year period. The disposal cost for this waste is estimated to be $16.2 million (1988 dollars) assuming a disposal fee at $40 per ton for surplus waste, and $45 per ton for the process rejects.

The total solid waste management cost for the third case is $70.6 million (1988 dollars). The total system cost includes the compost facility operation and the transportation and disposal of solid waste throughout the 20 year project period. The present value over the same 20 year period is $52.6 million assuming a five percent real interest rate.

Case IV

The last case discusses the economic effect and technical requirements associated with expanding the compost facility operations beyond the 8-hour processing shift. This case assumes the use of the Case I facility with the processing shift to be expanded from 8 to 10 hours per day.

This case is similar to the first one because it also uses the annual solid waste generation rate for the early 1990s. The 1990s generation rate of 46,300 tons per year represents the facility processing rate during the first nine years. It is also assumed that the facility will process 302 days per year (i.e., 6 days per week minus 10 planned holidays) and operate a single 8-hour shift. The resulting average daily processing rate would be 153 tons per day. During the remaining 11 years, the annual facility processing capacity is assumed to be 57,800 tons per year due to the two hour extension in the operating shift. During this 11 year period, the facility would process 302 days per year and achieve a daily processing rate at 191 tons per day.

The estimated capital investment is assumed identical to the first case. It must be noted that increasing the daily operating hours may require expanding the active composting and curing areas. The unit processing cost increase due to the capital investment required to expand the two processing areas can not be determined by the present costing methodology. The space requirements for the active composting and curing areas will depend upon: the throughput capacity of the facility; the desire to process solid waste during peak spring and fall generation periods; and the process system selected.

The total annual cost and the corresponding unit processing costs during the first nine years are also the same as the first case. However, the total operating costs are higher for the remaining 11 years. This would be due principally to additional operating and maintenance staff hours, extra electrical demand, and increased demand for spare parts and supplies. The tipping fee (not adjusted for inflation) is estimated at $44.70 per ton from 1991 to 1999 with a reduction to $42.40 per ton from the year 2000 to 2010.

In this case, the unit processing cost remains constant during the first nine years and decreases by 5 percent for the remaining years not accounting for inflation. The unit cost decreases during the last 11 years because the facility is operating 10 hours per day, rather than 8 hours. Because of the fixed capital investment, there is a reduction in the unit processing cost. The costs do not include the expense to transport and dispose the process rejects and surplus waste streams.

The following assumptions are used to estimate the change in the unit processing cost: Staff hours for the two additional processing hours will be increased by a factor of 1.5 to account for overtime pay requirements; The cost of electricity will increase in direct linear fashion as the processing hours increase; Maintenance, spare parts, and supplies will increase at a rate of 110% of the increase in processing hours. These assumptions result in an annual facility operating expense increase of $380,000 (1988 dollars) where the daily processing period increases from 8 to 10 hours.

During the twenty year project period, the facility will generate approximately 312,000 tons of process rejects and 113,000 tons of surplus waste, or a total of 425,000 tons of waste. The disposal cost for this waste is estimated to be $18.5 million (1988 dollars) assuming a disposal fee at $40 per ton for surplus waste, and $45 per ton for process rejects.

The total solid waste management cost for the fourth case is $64.1 million (1988 dollars). The total system cost includes the compost facility operation and the transportation and disposal of solid waste throughout the 20 year project period. The present value over the same 20 year period is $46.4 million assuming a five percent real interest rate.

Figure 4 graphically presents the solid waste diversion estimates for the four cases.

Comparison of the Four Cases

When the four cases are compared to each other, several important differences are identified. Referring to the information presented, the following observations can be made:

- The present value cost for waste disposal and operation of the proposed solid waste composting project increases from $45,400,000 to $52,600,000 as design capacity is increased and the amount of solid waste disposed at a sanitary landfill over the 20 year operating period decreases from 44 to 31 percent of the total waste stream.
- If the solid waste processing facility can increase the operating day from 8 to 10 or more hours, a 145 to 155 tpd facility can reduce the amount of unprocessed solid waste disposed at a sanitary landfill by 64 percent. Also, it would save capital investment facility costs by 10 to 19 percent; and reduce the unit processing costs by 2 to 9 percent through the 20 year operating period when compared to a straight 8-hour shift facility with the same daily capacity.
- If the facility is to operate on a standard 8-hour day and a solid waste landfill reduction of 60 to 65 percent is required, the facility should be sized to process 165 to 175 tpd (8 to 14 percent excess capacity) during the initial five to eight operating years, the facility unit processing cost would exceed $50 per ton and achieve the 60 to 65 percent landfill abatement throughout the 20 year period.

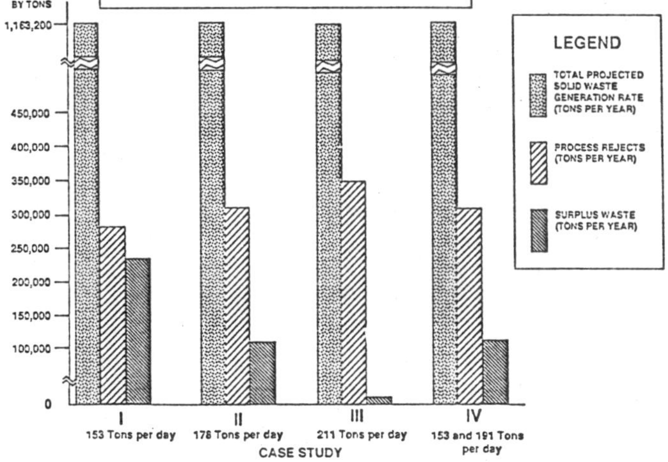

FIGURE 4: SOLID WASTE FLOW OVER THE 20 YEAR PROJECT PERIOD

Conclusion

Solid waste management policy goals associated with the amount of waste diverted from sanitary landfill disposal must be clearly defined and achievable. Once the goals are determined, the selection of a secondary material and solid waste compost processing train must be evaluated to assure the lowest possible tipping fee and cost of operation. The selection of a facility processing rate will have a direct impact on the amount of solid waste disposed at a sanitary landfill and the economics of the solid waste management program. During the facility design period, the selection of the processing capacity should be carefully considered. The results from the surplus waste/excess plant capacity computer model and present value calculations provide the owner/operator with tools to evaluate the facility processing rate with respect to economic and environmental concerns. Upon review of this information, the Wright County Project Management Team preferred that the facility be designed for expansion to a 10-hour processing day in the late-1990s. The initial MSW processing rate was selected at 165 tpd with future facility expansion to 205 tpd.

Mark Selby is a project engineer for DPRA Incorporated (formerly Pope-Reid Associates, Inc.), based in St. Paul, Minnesota.

Referemces

1. "Solid Wastes, Engineering Principles and Management Issues," George Tchobanoglous, 1977.
2. "Waste Management Planning, Evaluation, Technologies," David C. Wilson, 1981.
3. "Hennepin County Comprehensive Recycling Study," 1985.

38

"STARTERS"—INOCULUMS AND ENZYMES

Clarence G. Golueke and Luis F. Diaz

IN RECENT YEARS, a spate of stories has appeared in the popular press about "breakthroughs" in the discovery, isolation and "engineered" development of microbes capable of attacking refractory and hazardous substances hitherto deemed impregnable to biological degradation. One of the side effects of the stories was and is a mild revival of interest in the use of inoculums and other so-called starters and accelerators in the composting of wastes. This paper is presented in the hope of providing sufficient information about inoculums and enzymes to enable the reader to make a rational evaluation of the utility of starters and accelerators in the composting of sewage sludge, municipal solid waste (refuse), and animal manures.

An ability to make a rational evaluation is especially important, if for no other reason than to counteract a popular but ungrounded belief. The belief is that a given organism having been demonstrated effective under laboratory conditions and perhaps in small pilot-scale studies and "demonstrations" will be equally effective in conventional waste treatment practice. Overlooked is the fact that conditions in the laboratory, and to a considerable extent in pilot-scale operations, are maintained at optimum levels. The reality is that in conventional waste treatment practice, conditions rarely are uniformly optimum. Consequently, the potential abilities of the promising microorganisms fail to materialize.

This presentation begins with a short introduction to our terminology, and then proceeds to discuss the theories on which the use of starters in compost practice are based and the obstacles between theory and practice. In short, arguments are presented for and against the utility of starters. The discussion closes with: 1) A series of criteria and questions to be used in the evaluation of starters and the claims made for them; and 2), A few pointers on setting up an experiment to test the utility of a given starter on a particular substrate (waste).

Terminology

Usually, the term "starter" is given to the assortment of microbes and enzymes added to a particular substrate in the hope of 1) minimizing and even eliminating the initial lag period characteristic of biological processes; and 2), improving the compost process in general. (For our purposes, the substrate is the waste to be composted.) Although "starter" may be the term more commonly used, "accelerator" is probably more accurately descriptive, because shortening or eliminating the lag period would be an "acceleration" of the process.

As the term implies, the lag period is a time of very little obvious activity. In reality, however, it is the time required for indigenous and introduced microbiota to develop, adapt, and multiply to the level required for the succeeding periods of rapid proliferation and intensive activity. Nutrients and other substances used by the microbes in the lag period and in subsequent stages are obtained from the waste through the agency of enzymes synthesized by the organisms. The enzymes render the waste molecules amenable to microbial utilization. The net result is decomposition (composting) of waste. Usually, starters purveyed on the open market contain both microbes and their enzymes together with enzymes from other sources. However, from a microbiological point of view, it would seem that theoretically, only the enzymes would be needed.

Inoculation and Mass Inoculation

Reference to microbes in starters brings up the subject of inoculums and the importance of the quantitative distinction between "inoculation" and "mass inoculation." This distinction is particularly pertinent to biological treatment processes that involve wastes and to industrial fermentation.

As used in microbiology, "An inoculum is a *minute* sample removed from a pure culture of microbes for the purpose of starting a new culture or infecting an experimental animal."[1] (In waste management parlance, the qualification "pure culture" generally is ignored in favor of "culture of desirable (or desired) microorganisms.") The act of introducing the inoculum into the receiving medium is "inoculation." Unless otherwise specified, "inoculum" and "inoculation" usually refer in this paper to *small* or *minute* amounts as compared to the amount of the medium (substrate) to be inoculated.

"Mass inoculation" is the introduction of a *large* amount of inoculating material relative to the mass or volume of the substrate to be inoculated. In the fermentation and pharmaceutical industries, the amount is on the order of 10% of the substrate to be inoculated. Mass inoculation lends itself to batch and continuous types of systems. In batch operations, mass inoculation is accomplished by introducing (recycling) a portion of the previous output into the incoming raw waste. This last statement brings an intriguing thought to mind: If a particular inoculum should prove to be useful, recycling a part of the output should duplicate the benefits of the inoculum, because the output would contain representatives of the groups introduced by the inoculum.

Rationale

The rationale usually offered by proponents of inoculation is that it (inoculation) is a very useful adjunct in the biological treatment (especially composting) of sewage sludge, MSW, and other organic wastes. The utility is attributed to the supplying an abundant microbiota specifically developed to break down the particular waste. The addition purportedly facilitates the process and ensures a more complete breakdown of both the readily biodegradable and the refractory waste components. As will be shown, the reality is that all too frequently several factors intervene between the hoped-for utility and its materialization.

The primary advantage of mass inoculation over "ordinary" inoculation is the minimization or elimination of the lag period. Because a substantial fraction of the required microbial numbers is introduced, theoretically, process activity can begin without delay and the overall required retention period is shortened accordingly. The shortening permits a reduction in equipment and operational requirements. Another possible advantage over ordinary inoculation is that the sizable fraction of processed medium thus introduced may beneficially affect the composition and/or properties of the waste. Yet another advantage might be a lessening of the uncertainty associated with "ordinary" inoculation. It is not surprising, therefore, that mass inoculation rather than ordinary inoculation is the preferred approach for industrial applications. However, it must be remembered that to be of use, the microbial composition of the mass inoculum must be truly representative of the microbial complex needed for a successful outcome of the compost process.

Utility of Inoculation

Even though an inoculum may contain "engineered" microbes, to be useful in composting it must: 1) supply an essential type of microbe not already present in the raw waste; 2) augment a less than adequate microbial population; or 3) introduce a type or types of microbes more effective than those indigenous to the raw waste. Experience amply shows that the first and second conditions do not prevail in the wastes of concern in this paper, namely, sludge, MSW (including yard wastes and leaves), and manures. Exceptions due to failure to satisfy one or all of the three requirements could be wastes that have been sterilized or otherwise especially treated, or which come from certain processes (e.g., pharmaceutical waste products), or which are relatively homogeneous in composition (e.g., straw, xylem [wood]).[2]

Other than the preceding exceptions, demonstrating the absence of a specifically essential microbe from a particular mass of waste would be a monumental task. Two of the factors that underly the difficulty are: 1) The sizeable number of different groups of microbes involved, some of which may be more useful than others; and 2) Composting is a dynamic succession of microbial interactions with the substrate wastes. Because of these factors, it would be necessary to: 1) Isolate the active organisms from the waste; 2) Demonstrate the importance and degree of essentiality of the isolates to the composting process; 3) Identify the isolates, preferably following Koch's postulates (isolate, identify, inoculate, repetition of "symptoms"); and 4) Determine their respective roles in the compost process.

Determination of the roles often is attended by considerable uncertainty because of the difficulties encountered when dealing with mixed cultures. Nevertheless, a very practical, and hence extremely important reason for identifying the active types of microbes, relates to the maintenance and reproducibility of the particular inoculum. Obviously, it is necessary periodically to confirm the continued presence of desirable organisms in the stock culture and in subcultures from the stock culture. Reasons for the need are many. For example, due to the vicissitudes of the subculturing required to maintain a viable stock culture, the microbial composition of an inoculum has a tendency to change with time. Unfortunately, such changes rarely are for the better. The deterioration comes in part from a decline in virility (weakening) of microbial populations that usually attends repeated subculturing.

Finally, granted that the microorganisms were identified and their roles determined, the ability of the isolated microorganisms contained in an inoculum not only to compete with indigenous bacteria and but also to thrive in an external situation would be uncertain. The uncertainty is due in part to the aforesaid decline in virility.

Questions regarding the possibility or even the utility of attempting to introduce a type or types of microorganisms more effective than the indigenous microflora, remain to be answered. Regarding the possibility, it must be demonstrated that the microbe or microbes in the inoculum is or are truly more effective than the indigenous microorganisms. Although anecdotal accounts of successes due to the use of a particular inoculum are not unusual in the popular media, we have yet to come across unqualified accounts of successes in the refereed scientific and technical literature[3-6]. The degree of the credibility justified by reports in refereed scientific and technical literature or by any report that meets scientific criteria is far higher than that by anecdotal accounts in the non-refereed media. The reason is that with the former, we are assured that the reports are based on properly conducted scientific experiments and demonstrations; whereas with the latter, we have no such assurance. Scientific experimentation and demonstration minimize intentional and unintentional biases in the conduct of a demonstration and in the interpretation of its results. Unfortunately, the tendency is to rely upon anecdotal testimonials because the conduct of a truly scientific experiment or demonstration is not an easy task.

Granted that a truly effective inoculum were developed, the question remains as to its utility in compost practice. Many factors combine to negate its utility. A factor that immediately comes to mind is that the microorganisms inoculated into the waste are at a competitive disadvantage with respect to indigenous populations, which already are adapted to the conditions prevailing in the waste. By the time the introduced microbes do become adapted, the indigenous population will have increased to the limits permitted by the existing supply of nutrients. Such being the case, the added microbes would be of little use. A second factor is the dynamics of composting processes that in volve heterogeneous wastes such as sludge, MSW, and manures. The dynamics refer to the synchronization of the successions of microbes with changes in nature of substrate that result from the metabolic and assimilatory activities of the microbes. The synchronization involves a series of simultaneous expansions of microbial populations and breakdowns of substrate, all of which result in the successive appearance of groups of microbes adjusted to a stepwise reduction of complex substances to simpler compounds. Because of the synchronization, each group reaches its peak population at the time conditions have become optimum for it. Introduction of a particular group before that time (i.e., "ahead of its time") would be of little use, inasmuch as the group could not function at its peak capacity.

The aforesaid optimum conditions usually are nutritional in nature. The sequence of their occurrence is that of the removal of nutritional barriers posed by the complex composition of the substrate. Degree of complexity affects availability of elements (C, N, P, etc.) essential to a given type of microbe. Some microbes can utilize complex compounds, whereas others can assimilate only fairly simple ones. Some can only use inorganic compounds; and others, only organic compounds.

An excellent example of the stepwise requirement is found in the biological breakdown of certain halogenated compounds (some pesticides, PCBs)[7-9]. Research on the composting of halogenated compounds generally was unsuccessful until it was found that their biodegradation depended upon an anaerobic-aerobic sequence.

As "food for thought," we call attention to the claim made for some starters to the effect that they include microbes that can react with heavy metals to render them soluble, e.g., methylation of mercury and selenium. Such microbes do exist in nature, especially in the bottom mud of some streams. This capacity is indeed a "mixed blessing," inasmuch as an "inert" metal is rendered soluble and hence can be taken up by plants.[10-11]

We close our discussion on microbes by calling attention to certain opinions and conclusions expressed by Rozich, Colvin, and Gaudy in one of their papers.[12] Although their paper deals with hazardous wastes, it succinctly summarizes the message we endeavor to convey in our discussion. The three researchers point out that it is true that reliance on genetically modified microorganisms for enhancing biodegradation of toxic wastes may eventually yield results. However, many crucial microbiological, biochemical, and physiological questions must be addressed before applying these organisms to field treatment situations. An alternative to the genetic engineering approach is to utilize engineering parameters available for controlling a biological reactor to select desired population types. Given the diversity of natural populations, it should be possible to manipulate a reactor environment to maximize the rate of degradation of toxic organics. Practical questions to be answered deal with the identification of the selective environmental characteristics and a concomitant control strategy for maintaining the cultures under continuous flow conditions.

Enzymes

Most starter entrepreneurs include enzymes when listing the ingredients of their products. The background for this inclusion parallels the introduction of purportedly advanced versions of starters—i.e., "advanced" in terms of increased capacity, utility, and versatility. Thus, in the early 1950's (when the older of the authors made his appearance on the compost scene), starters were primarily microbial and references to identities of constituent microbes were very vague. References to enzymes were extremely few and far between. As early ("pioneer") researchers began to issue formal and informal reports on microbial groups (e.g., actinomycetes) observed by them, they also began to conjecture on the roles of those microbial groups in the compost process. The conjectures frequently were accompanied by surmises about the part played by enzymes.

Coincidentally, vendors of starters in vogue at the time

began to claim that their products included the newly reported microbial groups as well as an array of enzymes. For some reason, hormones were attracting attention at the time, and so most starters also were supposedly laced with hormones. In time, hormones began to disappear from the picture, whereas enzymes were given a billing parallel to that accorded the microbial component. The gain in the prominence given enzymes was not without foundation.[13-14] For example, workable processes were being developed for accomplishing enzymatic hydrolysis of cellulose—at first to produce a substrate for single-cell protein (e.g., yeast) production—and later, a sugar for ethanol fermentation.

In large part, the arguments regarding the utility of the microbial component of starters also apply to that of the enzymatic component. Thus, if the required microbes are present, their enzymatic output also will be present, because the synthesis of appropriate enzymes is an essential part of microbial activities in the compost process. However, it might be argued that the addition of certain enzymes not synthesized by the resident microbial population could well be beneficial, especially enzymes that have been "engineered" to break down refractory molecules.[13]

The difficulty in identifying the enzymes of most use in the composting process often surpasses that attending the identification of useful microbial groups. A substantial part of the difficulty is due to the high degree of the specificity of enzymes. To be of use, they must be matched with the substrates for which they are specific. The great strides made in the development of "engineered" enzymes have not eliminated the difficulty. It continues to be necessary to determine type and molecular structure of the enzyme required, changes that must be made in the enzyme's molecular structure, and ways of encoding the altered enzyme.

Not to be ignored when considering the addition of enzymes in a compost operation are the usually high cost of true enzymes, their sensitivity to environmental conditions, and their "fragility." The high cost is a function of the substantial expenses involved in producing enzymes—engineered or otherwise. Fragility is in part a function of sensitivity to environmental conditions. The sensitivity is keener than that of the microbes which synthesize them.

We close the "theoretical" portion of our paper by calling attention to an article which appeared recently.[15] Although it deals primarily with hazardous wastes and the verbiage is overly "popularized," it does convey much useful information that is both current and applicable to the subjects discussed in our paper.

Evaluating a Starter

This section has two parts. The first part is rather technical and presupposes some background in practical microbiology. In it are posed and discussed four questions and answers to be used in evaluating a starter. The second part is much less technical and presupposes only a broad knowledge and acquaintance with composting practice. In it is outlined a straightforward procedure for testing the utility of a specific starter for a particular waste.

Part 1: Questions and Answers:

Certain pieces of information are needed for an in-depth evaluation of a particular starter. The information can also be used in the selection of one starter in preference to another. The essence of this information is presented in the form of the following series of questions:

Question 1. Have the types or groups of organisms purportedly present in the starter or inoculum been demonstrated as being the same as those that are active in the composting process? An answer in the positive should be documented by a refereed publication. The document should answer the questions listed under question number 2.

Question 2. How were the organisms isolated, identified, etc.? Identity and descriptions of the microbiological techniques, procedures, etc. used in the isolation, etc.? Unless the answers are satisfactory, no assurance can be had that the organisms in the starter are the active ones or subcultures of the active organisms.

Question 3. How has the continuity of the culture been assured? Which procedures were followed for maintaining the culture? Continuity is an insurance that the organisms in the present starter are the same as those that were originally isolated. Maintenance procedure determines the "virility" of the organisms, i.e., ability of the organisms to function as expected and to compete with indigenous organisms. In other words, it has an important bearing on performance reproducibility. It is necessary to beware of mutations, of loss of organisms, and of contamination by "volunteer" organisms.

Question 4. This question deals with enzymes. How were the enzymes identified as being important to the process? The best recourse with regard to this question is to rely upon satisfactory documentation backed by professional advice on the interpretation of the documentation.

A serious problem with the four questions is that a substantial knowledge of microbiology and enzymology is required for an intelligent interpretation of the answers given them. On the other hand, such a need does not differ substantially from the need for the services of a professional in the design and implementation of a sizable compost operation.

Fortunately, this problem can be by-passed by the simple expedient of testing the starter or enzyme by way of the experimentation outlined in part 2. In fact, even though the four questions were satisfactorily answered and the answers were understood, it would be strongly advisable to test the starter on the waste to be composted and under the conditions to be encountered in the contemplated operation. To be worthwhile, such a test must take the form of an investigation based on sound scientific principles. Nevertheless, the test need not be difficult to set up and conduct.

Part 2. Testing a Starter:

We recommend that the turned windrow method be followed in the conduct of the test. Our reasoning is that conclusions based on a turned windrow test are applicable to other compost methods—including in-vessel composting. The conclusions can be validly extrapolated, because the basic principles of composting prevail regardless of method. Differences between methods as to performance arise from differences in ability to maintain composting conditions at optimum levels. Generally, the biggest differences are in rates. Provided retention times are adequate, differences in extent of breakdown and stabilization and characteristics of the product are minor.

Problems with using in-vessel composting for testing for the most part arise from "practicality." For example, an attempt to provide the essential requirement "simultaneity" would require the use of at least two identically designed, equipped, and operated vessels. Unfortunately, few operations are large enough to warrant two in-vessel units or are financially in a position to purchase more than one unit. However, even more important than practicality are certain basic impediments. Thus, granted that two vessels of a kind might be available, it is highly unlikely that both would function identically in every respect. On the other hand, testing by way of two sequential runs—one with and one without an inoculum, would introduce the possibility of transfer of key organisms from one run to the second, even though the control were the first in the sequence. After all, there is the possibility that the organisms indigenous to the waste might be the more useful. The reverse would be true if the starter were used in the first run of the series.

A useful, but easily run turned-pile test could be conducted as follows: Two equal portions are removed from a pile of well-mixed waste (intended substrate). The material in both portions must be representative of the material to be composted and must have been removed simultaneously. One of the two portions receives the starter, which is thoroughly incorporated into it. The other portion receives no starter, and is used to construct the "control" pile. The inoculated portion is used to construct the "variable" pile. The importance of having a "control" pile can not be over-stressed. A control is the *"sine qua non"* in the conduct of genuinely scientific investigation. Here, we emphasize that to qualify as a "control," *all* conditions for the control pile must *without* exception be precisely identical to those for the variable. Often overlooked are the nutrients and water contained in the medium in which the starter organisms are cultured or suspended. To avoid this oversight, the same volume of medium used to inoculate the variable should also be added to the control. Of course, the medium added to the control should be free of starter organisms.

Any one or combination of the conventional procedures may be followed in the construction of both piles. The key requisite is that with both piles, the procedures are *identical* and are applied simultaneously to the extent possible. Pile dimensions and geometry are those that are convenient to the situation. Obviously, they also must be identical. To eliminate environmental differences due to location of pile, the piles should be adjacent to each other.

As with the construction, conventional procedures (e.g., methods and frequency of turning, parameters, sampling) are followed in the conduct of the process. During the turning, especial care should be taken to minimize transfer of material from one pile to the other—i.e., avoid cross-inoculation. In closing our section on testing, we again emphasize that anything done to the variable pile must *also* be done to the control.

References

1. McGraw-Hill Dictionary of Scientific and Technological Terms, McGraw-Hill, Inc. New York, St. Louis, San Francisco, (1976)
2. Matthur, R.S., S.P. Magu, K.V. Sadasivam, and A.C. Gaur, "Accelerated Compost and Improved Yields." *BioCycle, 27*(2): 42–44 (Feb. 1986)
3. Solbraa, K., "An Analysis of Compost Starters Used on Spruce Bark." *BioCycle, 25* (2): 46, 47 (March 1984)
4. Michales, A., "Composting", *Public Works, 120* (1): 78 (Jan. 1989)
5. Farkasdi, G. "Do Additives Affect Windrow Composting of Refuse and Sludge?", *Comp. Sci., 6* (1): 11–13, (1966)
6. *Food, Fuel, and Fertilizers from Organic Wastes,* National Research Council, National Academy of Sciences, National Academy Press, Washington, D.C. (1981)
7. Quenson III, J.F., J.M. Tiedje, S.A. Boyd, "Reductive Dechlorination of Polychlorinated Biphenyls by Anaerobic Microorganisms from Sediments". *Science, 242*: 752–754 (4 Nov. 1989)
8. Vargas, C., R.C. Ahlert, E.E. Abbott, M.G. Gayton, "Anaerobic Degradation of Chlorinated Solvents", *Proceedings of 41st Industrial Waste Treatment Conference, May 1986,* Lewis Publishers, Inc., Chelsea, MI, (1987) pp. 339–346
9. Rojo, F., D.H. Pieper, K.,-H. Engesser, H.J. Knackmuss, K.N. Timmis, "Assemblage of Ortho Cleavage Route for Simultaneous Degradation of Chloro- and Methylaromatics", *Science, 238* (4832): 1395–1398 (4 Dec. 1987)
10. Kraut, J. "How Do Enzymes Work?" *Science, 242:* 533–539 (28 Oct. 1989)
11. Klibanov, A.M., "Enzymes: Nature's Chemical Machines", *Technology Review,* pp. 41–59 (Nov.-Dec. 1984)
12. Rozich, A.F., R.J. Colvin, A.F., Gaudy, Jr., "High-Rate Biological Process for Treatment of Phenolic Wastes", in *Biotechnology for Degradation of Toxic Chemicals in Hazardous Wastes,* ed. Scholze, Jr., *et al.* Noyes Data Corp., Park Ridge, NJ
13. Cappon, C.J., "Uptake and Speciation of Mercury and Selenium in Vegetables and Crops Grown on Compost-treated Soil", *Water, Air, and Soil Pollution, 34* (4): 353–361 (Aug. 1987)
14. Lee, T.A., J.K. Hardy, "Copper Uptake by the Water Hyacinth", *J. Environ. Health Sci.: Part A - Environ. Sci. & Engineering, A22* (2): 141–160 (1987)
15. "Will Bugs Eat Up the Cleanup Market?", *Waste Age, 19* (10): 60, 61, 64, 66, 68, 70 (Oct. 1988)

39

LANDSPREADING MUNICIPAL WASTE COMPOST

D.A. Mays and P.M. Giordano

WIDESPREAD INTEREST in composting municipal solid waste first developed in the United States about 25 years ago. The driving force behind composting at that time was the hope of selling the finished product at a profit. There was not yet a widespread interest in resource conservation and recycling or the difficulty and cost in siting and developing landfills which exists today. Among early examples of composting ventures were those located at Altoona, Pennsylvania; St. Petersburg, Florida; and Mobile, Alabama. Given the low landfill costs of that era and the general lack of recycling technology and opportunities, most if not all, composting ventures proved to be unprofitable.

In the latter half of the 1960s, the Tennessee Valley Authority entered into a cooperative agreement with the U.S. Public Health Service (a forerunner of EPA) and Johnson City, Tennessee, to build and operate a demonstration-scale municipal solid waste composting plant. This facility was located on one of the wastewater treatment plant sites near Johnson City in northeast Tennessee.

The processing was somewhat crude by today's standards in that not much except white goods and ferrous metals was removed from the trash prior to grinding. After grinding, about 25% by volume of sewage sludge was added to increase the nutrient value and the resulting mixture was windrow composted for several weeks with periodic turning. Partial analysis of a typical batch of finished compost is shown in Table 1.

The finished compost was used in a large number of crop production and surface-mine reclamation demonstrations in east Tennessee and western North Carolina. Replicated crop production experiments using compost as a soil amendment were conducted at Johnson City and at Muscle Shoals, Alabama.

Experimental Procedure

Compost was applied on the Johnson City field plots annually from 1968 through 1972 at rates ranging from 8 to 200 tons/acre, resulting in total applications ranging from 40 to 1,000 tons/acre. A small amount of supplemental fertilizer supplying N, P, and K at rates of 50, 22, and 42 lbs/acre, respectively, was applied each year. Field corn was grown annually from 1969 through 1983 using conventional soil tillage techniques. Atrazine was applied preemergence for weed control. Corn grain yields were measured each year.

In October 1987, selected plots within the experimental site were soil sampled to a depth of 36 inches by 6-inch increments in order to document the long-term effects of compost application on various chemical characteristics of the soil. Four 2-inch diameter soil cores were taken from each plot. Soil cores from the four replicated plots were composited by depth prior to analysis. Plant nutrient, pH, and cation exchange capacity determinations were performed by the Auburn University soil testing laboratory while the organic matter and heavy metal analyses were performed in the TVA laboratories.

TABLE 1.

Typical analysis of compost produced at Johnson City, Tennessee.

Element	Concentration, %	Element	Concentration, ppm
C	19.10	Zn	1500
N	1.26	Cu	300
P	0.44	Mn	500
K	0.52	Cd	20
Mg	0.49		
Ca	4.70		
S	0.40		
Na	0.42		

Successful Corn Crops

Thirteen successful corn crops were produced over a 15-year period. Yield increases due to compost application were routinely noted throughout the 14 years (Fig. 1). Total yield increase over that period ranged from 55% with a total application of 40 tons/acre to 153% where the total compost application was 1,000 tons/acre.

Physical evaluations of soil done at various times in the life of the experiment show that compost applied at high rates decreased bulk density and resistance to penetration and increased water holding capacity. The 1987 soil sampling and analysis were done to document the long-term chemical changes resulting from compost usage and to determine if it appeared that environmental hazards had been created.

Because the compost had a high calcium content (4.7%), it exerted a pronounced long-term liming effect on the soil. Even the low total applications of 40 and 80 tons/acre maintained the pH in the top 18 inches of soil above 6.1 for 14 years after the last applications. Rates of 250 tons/acre and above maintained the soil pH above 7.0 in the top 18 inches and above 6.5 to a depth of 24 inches. In contrast, the untreated soil had a pH of 5.6 at the surface and 4.7 at the 24-inch depth.

FIGURE 1.

Effect of municipal compost application on corn yield.

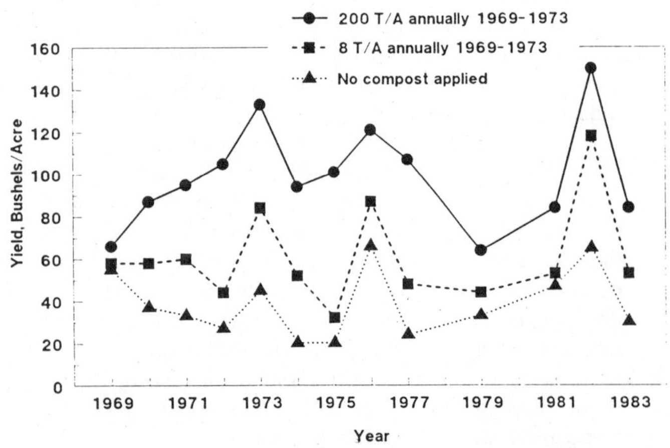

Cation exchange capacity and soil organic matter in the upper layer of soil were both increased significantly at application rates of 250 tons/acre and higher (Table 2). Since these changes were not noted below the 12-inch depth, it must be assumed that compost mixing during soil tillage was the only mode of downward movement of organic matter and its associated effect on cation exchange capacity.

The primary and secondary plant nutrient levels were all increased significantly because of the large amounts supplied by the compost, but the amount of downward movement varied with the nutrient (data not shown). Increased phosphorus levels were found only in the zone of physical mixing (above 12 inches). Potassium levels in the soil increased down to the 36-inch depth. Calcium levels increased with increasing compost rates to the 30-inch depth. Magnesium levels were increased in the upper 12 inches, but varied considerably below that, suggesting that magnesium ions may have been displaced from the exchange sites and leached from the profile where large amounts of calcium and potassium ions were present.

Analysis of the components of the compost indicated that most of the heavy metal load was supplied by the sewage sludge. It should be noted that there was little con-

TABLE 2.

Effect of municipal compost on cation exchange capacity and soil organic matter concentration.

Total compost application, tons/Ac	0″-6″ depth		6″-12″ depth	
	CEC	O.M., %	CEC	O.M., %
0	6.9	1.6	6.2	1.0
40	7.4	1.7	6.9	0.8
80	7.0	1.7	6.6	1.0
250	10.5	2.4	9.4	1.4
500	10.8	3.6	10.7	2.0
1000	14.6	4.9	10.0	2.3

TABLE 3.
Effect of municipal compost on heavy metal concentration in soil.

Total compost application, tons/Ac	0″-6″ depth				6″-12″ depth				12″-18″ depth			
	Cd	Cu	Pb	Ni	Cd	Cu	Pb	Ni	Cd	Cu	Pb	Ni
0	0.08	2.1	3.2	0.14	0.04	1.3	2.0	0.06	0.00	0.3	0.8	0.03
40	0.17	4.7	7.1	0.19	0.11	3.2	4.7	0.12	0.02	0.5	0.9	0.02
80	0.22	5.1	7.8	0.18	0.09	2.9	4.2	0.10	0.00	0.4	0.8	0.03
250	0.39	9.3	14.1	0.29	0.25	6.3	8.9	0.26	0.02	0.6	0.9	0.04
500	0.62	16.3	21.2	0.55	0.40	14.7	13.3	0.49	0.02	0.6	0.8	0.13
1000	1.01	26.6	34.3	1.02	0.30	10.0	10.3	0.36	0.01	0.8	0.9	0.08

cern about heavy metals at that time and pretreatment was not a condition for having industrial wastes accepted by wastewater treatment plants. The data in Table 3 show that, while heavy metal loads in the surface soil were correlated with the amount of compost applied, there was virtually no movement of metals below the 12-inch depth.

As with phosphorus and organic matters, it is likely that most of the movement into the 6- to 12-inch layer resulted from physical mixing during tillage. These data support the conclusion that there is likely no hazard of groundwater contamination by heavy metals applied in sludge or compost. The 19 years since compost application began would seem to be ample time for downward movement to the 12- to 18-inch layer to begin if it were going to happen.

In conclusion, the data collected over a 19-year period show that the beneficial effects of municipal compost application include sustained higher crop yields, a more favorable soil pH, increased organic matter and cation exchange capacity, and enhanced supplies of primary and secondary plant nutrients. While heavy metals were increased in the surface soil by this particular compost, there was no evidence of downward movement which could routinely contaminate groundwater.

David Mays was formerly with the Department of Agricultural Research, Tennessee Valley Authority, in Muscle Shoals, Alabama, and is currently with B&H Environmental Services of Hermitage, Tennessee. Paul Giordano is Manager of the Agricultural Research Department, TVA.

Reference

Composting at Johnson City—Final Report on Joint U.S. EPA-TVA Project. EPA/530/SW-317.2. November 1975.

40

THE NATIONAL MOOD AND STATUS OF MSW COMPOSTING

Clarence G. Golueke and Luis F. Diaz

THREE DECADES of first-hand observation of the vicissitudes of composting as a waste management option have demonstrated to the authors the important bearing of the national mood on the status of composting. Paradoxically, the national mood simultaneously reflects and yet fundamentally establishes national priorities. If the mood is "upbeat" about a particular movement, it makes the success of that movement a top priority. A high priority brings with it the willingness to do whatever is needed to ensure the successful course of the movement. Illustrating the bearing of the national mood upon the course of the compost "movement" is the aim of this report.

The national mood is alternately characterized by optimism, by pessimism, or simply by indifference. Optimism brings with it a willingness and even eagerness to do something about a situation. Conversely, pessimism and indifference bring inaction or even opposition.

As we see it, the past three and a half decades began with a mood redolent of complacency. After a decade or so, complacency began to fade to concern and anxiety (transition mood), which in turn, flared into an urgency best described as an unlikely combination of desperation (pessimism) and optimism. This report is divided into three main parts on the basis of those three moods. Within the time span-covered, the mood of complacency extended from 1950 to the mid-1960's ("early days"); the transition mood, from the mid-1960's to the mid-1970's; and the mood of urgency, from the mid-1970's to the present. (These time spans are approximations.)

Part I. Mood of Complacency

The complacency of the early days was largely a function of a plethora of misconceptions and ungrounded convictions. A prevailing conviction downplayed the importance of fresh and decomposing MSW in terms of its seriously constituting a threat to the quality of the environment and to that of the air, land, and water resources and to private and public health. At the most, it was admitted that the waste might be a nuisance to any one who happened to come in contact or be in very close proximity to it. Vector breeding and proliferation were regarded as merely being temporary nuisances. Moreover, the popular feeling was that if any problems were to arise, they would be easily solved by science and technology, both of which were thought to "have no limits as far as finding solutions are concerned."

The prevailing lack of concern was reflected by the minimum of effort allotted to the management of municipal waste. Symptomatic of the unconcern was the prevalence of the open dump in MSW disposal. As the decade wore on and the amounts of MSW began to reach unmanageable proportions, burial began to be regarded as the "way to go", and a primitive version of the "sanitary landfill" evolved. This early version of the sanitary landfill was considered as being the solution to all MSW disposal problems. The technology was straightforward—dig a hole, deposit the wastes, compact them and cover them with a layer of dirt, and plant the cover with vegetation in the hope of placating the environmentalists. Plenty of vacant land was available. Moreover, there were those abandoned quarries, gullies, canyons, and swamps and tidal flats to be "reclaimed". thus, under the euphemism, "reclamation", unconscionably large acreages of the San Francisco Bay tidal flats became the depositories for metropolitan wastes. The point of importance here is that with the exception of a few individuals and groups popularly classed as "over-zealous and misguided environmentalists and conservationists", the popular conviction was that disposal in or on the land is the most satisfactory final destiny for all urban wastes. This hopelessly naive view arose from two misconceptions: 1) A well-nigh infinite supply of land is available. 2) Biological and physical processes in the soil surrounding the fill would convert objectionable substances in landfill leachate to innocuous forms ("attenuation") before the leachate could reach the groundwater. This attenuation would render leachate innocuous with respect to the water resource.

An important fact to remember is that even in its then primitive form, sanitary landfilling was far from being widely practiced. On the contrary, the open dump continued to be the popular means of waste disposal well into the 1970's.

The upshot of all of this naivete was that composting was popularly regarded as being a luxury attractive only to a few pesky activists. This situation and the very real disastrous financial outcomes of several composting ventures plus the undeniable environmental insults generated by an occasional poorly managed compost operation combined to make the climate for composting exceedingly unfavorable. The climate was not improved by the panacean attribute accorded to composting by some overly-enthusiastic compost proponents with respect to its being the ultimate solution to all waste disposal problems. It did not help that the financial disasters were due to a number of causes, of which naivete and gullibility may be numbered among the primary factors; or that the unfortunate environmental insults could be explained by the poor design of some plants and the mismanagement of others.

In passing, it might be mentioned that during this period, incineration also fared poorly in terms of acceptance. The few operations that were started, met with untimely ends.

Part II. Rise of Concern and Anxiety: Transition Mood

The well-remembered "activist decade" ran its course during the transition period. Both events were marked by a growing sense of anxiety and desperation triggered by the exposure of old shibboleths and gradual loss of naivete. The exposure and loss of naivete combined to erode national complacency and replace it with concern and anxiety. This development not only affected political and social circles, it also fundamentally altered the public view of the environment, in that it brought an awareness of the paramount importance of the relation between the environment and the quality of life and the public well-being.

Furthermore, it brought into sharp focus, the limitations of nature's ability to heal itself and of science and technology to come up with sure-fire solutions. In addition, the former conviction that our natural resources were inexhaustible and that any shortfalls could be compensated by technological advances, began to be shaken. A corollary was the broadening recognition of the need for resource conservation. To the public at large, "environmentalist" and "conservationist" ceased being epithets. It was inevitable that all of these social and political developments would impinge upon waste management.

The onset of serious concerns and doubts about the role of landfilling in solid waste disposal marked the beginning of a revival of the dormant interest in composting. These doubts and concerns are readily traceable to the urban sprawl that was and is making landfill siting increasingly difficult and frustrating. A host of factors conspired to overwhelm landfilling with an avalanche of restrictive design, operational regulations and legal specifications. The outcome was and is a sharp escalation of capital, operational and maintenance costs. Compounding the problem is the fact that responsibility no longer ends with the completion of the fill. On the contrary, it now has been made to continue for some time thereafter.

Incineration, hitherto a rather minor element in waste disposal practice (i.e., in comparison with landfilling), began to be considered as a likely successor to landfilling for primacy in waste disposal. Incineration seemingly had many things in its favor, such as a small land area requirement, complete containment of the wastes, and a sharp reduction in volume of waste, i.e., volume of residue was but a small fraction of that of the raw waste. Unfortunately, incineration had several inherent characteristics that unless modified or controlled would have a deteriorating effect on the environment and the public health. The problem is that bringing about the control is difficult and expensive. As a result, the suitability of incineration was seriously questioned, and its acceptance as a disposal practice was very restrained during the transition period.

The unfavorable developments in landfilling and incineration led to a sharp improvement in the prospects of composting. The improvement was accentuated by the popular recognition of the conservational nature of composting. Unlike landfilling and incineration, composting reclaims and recycles resources rather than bury or destroy them. However, as far as adoption as a waste management practice is concerned, its primary attraction is its inherent compatibility with the environment and the public well being. Its conservational potential is regarded as a benefit.

An exceedingly important development was the deserved recognition of the attractive features of composting as a mean of treating sewage sludge. Background factors leading to the recognition were the vexatious problems that beset the landfilling and the incineration of sludge soon after the output of sludge began to burgeon in the 1970's. Consequently, alternatives to the two methods were urgently sought. The suitability of composting not only as a treatment device but also as means of recovering useful plant nutrients made it (composting) an attractive option. Fortunately, careful research augmented by adversity had gone far towards removing naivete and promoting realistic attitudes concerning composting. Although it was true that the advances still had a long way to go before reaching satisfactory levels, progress was sufficiently far to permit composting to be ranked as a leading method of sludge treatment and disposal.

Unfortunately, progress of composting in municipal waste treatment was much less impressive. Although the panacean attributes and exaggerated claims so abundant in the preceding era had been muted, they continued to depart from realism. All too many of the breakthroughs being touted for compost equipment and system design proved to be recycled versions of earlier designs and equipment. However, the biggest difficulty was in preprocessing the incoming waste stream such that a substrate amenable for composting could be produced. The required research and development regarding design and manufacture of needed preprocessing equipment was only beginning. In fact, the investigation needed for preprocessing itself was only beginning.

Part III. Mood of Urgency

The desperation aspect of the national mood of urgency is an accompaniment of the dawning realization that our natural resources and with them the environment and that greatest of all resources, the public health and well-being, are being rapidly degraded and in some cases depleted to the point of no return. Because of the intimate association between waste disposal and the land, air and water resources, the environment, and private and public well-being, that realization exerts a direct effect on waste management and hence, on composting. Accordingly, the mood demands that wastes be managed such that the association not jeopardize resources, environment and health. Composting fits the mood very well because it (composting) is a means not only of rendering the organic fraction of wastes harmless, but also of reclaiming useful plant fertilizer elements and enriching the land resource by returning organic matter to it.

Unfortunately, until a few years ago, being well suited to the national mood was not enough to give composting the impetus needed for making inroads on the landfill-incineration competition. Since then, however, the tables have been turned. The turnabout was triggered by the factors discussed in the section that follows.

Factors Responsible for the Turn-About:

Scarcity of Land: The growing scarcity of arable land for satisfying agricultural demands together with the much-publicized, ever-expanding urban and industrial sprawl have transformed the status of the shortage of land for waste disposal from threatened to existing or actual. In the competition for the highly limited land area as yet available, demands generated by urban-industrial sprawl are given priority over those for landfilling waste. It is true that as yet there is an abundance of distant sparsely populated open areas, which for many reasons are not likely to be urbanized in the foreseeable future, and which are not suited to agriculture. The problem is that transportation costs, insistence upon the preservation of pristine environment, and reluctance to accept imported wastes are likely to make the availability of those lands largely illusory. Moreover, it is becoming increasingly apparent that the consequences of adverse impacts exerted on seemingly distant lands are all too often felt far beyond those lands. The upshot is that the prospects for landfilling are not as favorable as they were formerly. It should be emphasized, however, that landfilling, however curtailed, will always be with us.

High Costs of Sanitary Landfilling: Scarcity of land and the resulting potentially high transportation costs are not the only negative factors in the status of landfilling. In addition, there are the substantial expenditures involved in sufficiently isolating the buried wastes and their breakdown products from the environment. Isolation involves more than burial of the raw wastes, it extends to the complete containment of the decomposition products of the wastes.

Problems of Incineration: As is the case with incineration of sewage sludge, incineration of municipal wastes is beset with many problems, not the least of which are the escalating costs of the air pollution control demanded by the various governmental regulatory bodies. The escalation is a result of the necessity for trapping the growing number of toxic substances found in incineration emissions. The difficulty is compounded by the ongoing dispute about the disposal of the ash residue. Unfortunately, the body of evidence in favor of classifying the ash as a toxic substance is growing.

Conclusions: In summary, the constantly rising costs of landfilling and incineration together with increasingly stricter regulations and higher standards are causing the hitherto lopsided competition between the landfill-incineration combination and composting to lean heavily in favor of composting. The position of composting is further improved by the fact that as a practice, it is more in keeping with the prevailing pro-environment national mood than are landfilling and incineration. Thus, the favorable set of circumstances that propelled sludge composting into widespread acceptance and practice is now ready to do the same for MSW composting. The only requirement is that the favorable set of circumstances not be thwarted through technological and economical naivete and uncritical optimism. This requirement can be met by keeping certain fundamental facts and precautions in mind.

Fundamental Facts and Precautions: Not surprisingly, the waste treatment potential of composting is one of the fundamental factors that apparently have been most influential in determining its degree of acceptance in waste management practice. A decisive factor in evaluating this potential was and is economics, i.e. as compared with those of sanitary landfilling and incineration. Thus, when it was found that sludge could be satisfactorily composted at a cost comparable to that of incineration and with a far less environmental insult, composting soon began to displace incineration in sludge disposal. Moreover, this displacement continues despite the increasing number of constraints placed on the use of the compost product in agriculture. The point here is that in sewage sludge management, the superiority of composting as a waste treatment process is the key factor in its acceptance.

That economics is as important to the acceptance of MSW composting as it is for sewage sludge composting, is evidenced by the past record of the process in MSW management. The utility of composting as a highly successful resource conservation device was recognized from the earliest days of its application. Yet, despite that recognition, MSW composting never gained the acceptance held by landfilling and incineration. The major reason for the failure was that even though the desirability of conservation was recognized, economics remained the decisive factor in favor of sanitary landfilling and incineration.

Stressing the importance of economics should not be construed as downplaying the value of composting's conservation potential. Indeed, recent developments are showing that when the economic requirements and the satisfaction of performance of composting are the equal of those of the competitive processes, then the conservation potential could be the decisive factor in favor of composting. Morreover, gaseous emissions and water pollution continue to be important factors in incineration and landfilling.

We close this paper with a few suggestions:

1. Do not overburden the economics of composting with adding unnecessary steps or by resorting to needless sophistication. (An extreme but true example: A design that calls for passing the waste through two successive reactors—Metro-type reactor followed by a drum-type reactor.)
2. Do not abbreviate preprocessing. The temptation to abbreviate preprocessing arises when costs are higher than anticipated. We are aware of two ongoing operations in which preprocessing is being ignored. Their compost products are thoroughly unacceptable.
3. System design and implementation should be left to professionals qualified for compost applications through experience and demonstrated competence.
4. Be thoroughly critical, if not downrightly skeptical, when evaluating claims made by proponents of particular systems. Seek the advice and ascertain the opinions of qualified, objective experts.

APPENDIX

DIRECTORY

41

COMPOSTING EQUIPMENT AND SYSTEMS

COMPOSTING SYSTEMS: SLUDGE, MSW, YARD WASTE

AD+SOIL
1151 Pocopson Road
West Chester, PA 19382
215-793-3200

Agripost, Inc.
PO Box 2449
Pompano Beach, FL 33061
305-971-9100

American Bio Tech, Inc.
PO Box 19769
Jacksonville, FL 32245
904-724-2424

American Recovery Corp.
2000 Pennsylvania Ave., N.W.
Suite 8350
Washington, D.C. 20006
202-457-6610

Ashbrook-Simon-Hartley
PO Box 16327
Houston, TX 77222
713-449-0322

Bedminster Bioconversion Corp.
52 Haddonfield-Berlin Road
Suite 1000
Cherry Hill, NJ 08034
609-795-5767

Buhler-Miag, Inc.
PO Box 9497
Minneapolis, MN 55440—9497
612-545-1401

California Co-Composting Systems, Inc.
PO Box 10998
Marina del Rey, CA 90295
213-306-4195

Cholla Waste Management, Inc.
114 West 7th Street
500 Norwood Tower
Austin, TX 78701
512-482-0806

Cobey Composting
4250 S.R. 309
Galion, OH 44833
419-468-2288

Compost Systems Co.
9403 Kenwood Road
Cincinnati, OH 45242
513-984-0625

Cooperative Waste Industries, Inc.
271 LaFond Avenue
St. Paul, MN 55103
612-488-9246

Daneco, Inc.
450 Park Avenue
Suite 2104
New York, NY 10022
212-582-8150

Ebara Environmental Corp.
PO Box 516
Greensburg, PA 15601
412-832-1200

Ecological Technologies, Inc.
307 W. 7th Street
Suite 1400
Ft. Worth, TX 76102
800-634-8370

Enviro-Gro Technologies, Inc.
3700 Koppers Street
Baltimore, MD 21227
301-644-9600

Environmental Recovery Systems, Inc.
1625 Broadway, #2600
Denver, CO 80202
303-623-1011

Estech Corporation
1907 American Drive

Neenah, WI 54956
414-725-6555

Fairfield Service Co.
240 Boone Avenue
Marion, OH 43302
614-387-3335

Farmer Automatic
PO Box 39
Register, GA 30452
912-681-2763

Geotech Industries, Inc.
1019 W. Dixie Avenue
Leesburg, FL 32748
904-728-2245

Harbert/Triga
PO Box 1297
Birmingham, AL 35207
205-985-5403

International Process Systems
PO Box 143
Lebanon, CT 06249
203-642-6670

Lundell Manufacturing Co.
PO Box 171
Cherokee, IA 51012
712-722-3709

Omni Technical Services, Inc.
50 Charles Lindberg Blvd.
Uniondale, NY 11553
516-222-0708

Organic Waste Systems
Dok Noord 7b
B-9000 Gent, Belgium
32/91/33.02.04

OTV
11 av Dubonnet
92407 Courbevoie Cedex
France
1 49 04 46 64

PLM Sellberg
c/o Waste Management, Inc.
3003 Butterfield Rd.
Oak Brook, IL 60521
312-572-8800

Purac Engineering, Inc.
Commonwealth Building
University Office Plaza
Newark, DE 19702
302-737-3355

Recomp, Inc.
1500 E. 79th Street
Suite 102
Bloomington, MN 55420
612-854-6211

Renewable Resource Recovery
8105 Perry Highway
Pittsburgh, PA
412-366-4260

Reuter, Inc.
410 11th Avenue, South
Hopkins, MN 55343-7878
612-935-6921

Riedel Waste Disposal Systems, Inc.
PO Box 5007
Portland, OR 97208
503-240-8692

Royer Industries, Inc.
PO Box 1232
Kingston, PA 18704
717-287-9624

Taulman, Inc.
415 E. Paces Ferry Rd., N.E.
Atlanta, GA 30305
404-262-3131

Trash Reduction Systems, Inc.
10901 Lowell, Suite 125
Overland Park, KS 66210
913-661-9494

Valorga Systems
BP 56
34740 Vendargues
France
(33) 67 87 80 00

Voest-Alpine Composting System
c/o Chambers Development Co., Inc.
10700 Frankstown Rd.
Pittsburgh, PA 15235
412-242-6237

WPF Corporation
PO Box 381
Bellevue, OH 44811
419-483-7752

Waste Processing Corp.
One Appletree Sq., Suite 961
Bloomington, MN 55425
612-854-8666

CONVEYORS

American Pulverizer
5540 W. Park Avenue
St. Louis, MO 63110
314-781-6100

AR Products
PO Box 703
Indian Trail, NC 28079
704-821-6342

ATARA Incorporated
1114 Goffle Road
Hawthorne, NJ 07506
201-423-9411

Athey Products Corp.
PO Box 669
Raleigh, NC 27602
919-556-5171

C-Tech Conveyor Corp.
PO Box 1068
Branford, CT 06405
203-481-2141

Carrier Equipment, Inc.
PO Box 37070
Louisville, KY 40233
502-969-3171

Dover Conveyor
PO Box 300
Midvale, OH 44653
614-922-9390

Esco Corporation
2141 N.W. 25th Ave.
Portland, OR 92710
503-228-2141

Hallco Mfg. Co.
PO Box 20519
Seattle, WA 98102
206-282-0653

Heil Engineered Systems
PO Box 15877
Milwaukee, WI 53215
414-647-3439

Hobbs-Adams Engineering
1100 Holland Road
Suffolk, VA 23434
804-539-0231

Hustler Conveyor Co.
4985 Fyler Avenue
St. Louis, MO 63139
314-352-6000

Keith Manufacturing Co.
PO Box 1
Madras, OR 97741
503-475-3802

Lummus Development Corp.
PO Box 2526
Columbus, GA 31902
404-323-1081

Mayfran International
6650 Beta Drive
Mayfield, OH 44143
216-461-4100

Nott Company
PO Box 1422
Minneapolis, MN 55440
612-781-9561

Prab Conveyors, Inc.
5944 E. Kilgore Road
Kalamazoo, MI 49001
616-349-8761

Rexnord, Inc.
PO Box 2022
Milwaukee, WI 53201
414-643-3000

Selco Products, Inc.
PO Box 406
Baxley, GA 31513
912-367-4661

Serpentix Conveyor Corp.
9085 Marshall Ct.
Westminster, CO 80030
303-430-8427

Jervis B. Webb Co.
Webb Drive
Farmington Hills, MI 48018
313-553-1220

Webb Materials Handling Equipment
560 Industrial Drive, N.E.
Marietta, GA 30062

Wolf & Associates
4860 Mustang Circle
St. Paul, MN 55112
612-780-4550

Universal Engineering Corp.
800 1st Avenue, N.W.
Cedar Rapids, IA 52405
319-365-0441

DEWATERING SYSTEMS

AFL Industries, Inc.
3661 W. Blue Heron Blvd.
Riviera Beach, FL 33404
305-844-5200

Alar Engineering Corp.
9651 W. 196th St.
Mokena, IL 60488
312-479-6100

Arus-Andritz
Arlington S. Industrial Park
1010 Commercial Blvd.
South Arlington, TX 76017
817-465-5611

Ashbrook-Simon-Hartley
11600 E. Hardy
Houston, TX 77093
713-449-0322

Bio Gro Systems
PO Box 209
Annapolis, MD 21404
301-263-2237

Bio-Nomic Resources
1101 Richmond Ave.
Pt. Pleasant Beach, NJ 08742
201-295-3551

Brown Bear Corp.
PO Box 148
Lenox, IA 50851
515-333-4551

The Carylon Group
1000 South Western Ave.
Chicago, IL 60612-4138
800-621-4342
800-621-2119 (IL only)

Consolidated Sludge
670 S. Green Rd.
Cleveland, OH 44121
216-381-5371

Enviroquip, Inc.
2526 Kramer Lane
PO Box 9069
Austin, TX 78766
512-836-1614

Envirex
PO Box 1067
Waukesha, WI 53187
414-547-0141

Hydropress, Inc.
59 Dwight St.
Hatfield, MA 01038
413-247-9656

Infilco Degremont, Inc.
PO Box 29599
Richmond, VA 23229
804-281-7600

J.D. Meagher & Co.
PO Box 752
Westborough, MA 01581
617-366-6606

Jones Environmental
1931 Industrial Dr.
Libertyville, IL 60048
312-367-5440

K-F Environmental Technologies, Inc.
17 Furler St.
Totowa, NJ 07512
201-256-6226

Komline-Sanderson Engineering Corp.
12 Holland Ave.
Peapack, NJ 07977
201-234-1000

Mobile Dredging & Pumping Co.
344 Pottstown Rd.
Exton, PA 19341
215-363-6677

National Dredging & Pumping Corp.
Box 127
Folcroft, PA 19032
215-237-0700

Parkson Corp.
PO Box 408399
Ft. Lauderdale, FL 33340
305-974-6610

Plesh Industries, Inc.
One River Rock Drive
Buffalo, NY 14207
716-873-4916

Roediger Pittsburgh, Inc.
R.J. Casey Industrial Park
Columbus & Preble Avenues
Pittsburgh, PA 15233
412-2331-7979

Somat, Inc.
Pomeroy, PA 19367-9990
215-384-7000

D.R. Sperry & Co.
112 North Grant Street
North Aurora, IL 60542

Stord Bartz Americas, Inc.
309 Regional Rd., S.
Greensboro, SC 27409
919-668-7727

Tricil Environmental Response
PO Box 19529
Houston, TX 77224
713-467-3433

MAGNETIC SEPARATION SYSTEMS

Applied Magnetic Systems, Inc.
PO Box 20911
Greenfield, WI 53220
414-321-9739

Dings Magnetic Co.
4740 W. Electric Ave.
Milwaukee, WI 53219
414-672-7830

Eriez Magnetics
Asbury Road at Airport
PO Box 10608
Erie, PA 16514
814-833-9881

Newell Manufacturing Co.
PO Box 9367
San Antonio, TX 78204
512-227-9090

Stearns Magnetics, Inc.
6001 S. General Ave.
Cudahy, WI 53110
414-769-8000

Walker/Bux-Shrader
15142 Vista Del Rio

Chino, CA 91710
714-597-4785

MARKETING

American Soil Products
2222 Third St.
Berkeley, CA 94710
415-540-8011

Compost Management, Inc.
354 Main St., N.
Doylestown, PA 18901
215-348-9288

Kellogg Supply Co.
350 W. Sepulveda Blvd.
Carson, CA 90745
213-830-2206

Soil Products, Inc.
PO Box 145
Hermitage, TN 37076
615-889-4091

MATERIALS HANDLING

Ag-Bag Corp.
PO Box 418
Astoria, OR 97103

Aggregates Equipment, Inc.
PO Box 39
Leola, PA 17540
717-656-2131

Bio Gro Systems
PO Box 209
Annapolis, MD 21404
391-263-2237

Buhler-Miag, Inc.
PO Box 9497
Minneapolis, MN 55440
612-545-1401

Converto Mfg. Inc.
PO Box 287
Cambridge City, IN 47327
317-478-3205

East Mfg. Co.
PO Box 277
Randolph, OH 44265
216-325-9921

Eastern Conveyor Co., Inc.
4345 Orchard St.,
PO Box 9584
Philadelphia, PA 19124
215-289-7585

Kolman/Athey
Box 806
Sioux Falls, SD 57101
605-336-2610

Laidig, Inc.
14535 Dragoon Trail
Mishawaka, IN 46544
219-256-0204

Layco Mfg., Inc.
RR #2, PO Box 100
Marshall, IL 62441
217-826-6352

Nott Co.
PO Box 1422
Minneapolis, MN 55440
612-781-9561

Phelps Industries, Inc.
PO Box 1093
Little Rock, AR 72203
501-375-1141

Retec, Inc.
PO Box 724634
Atlanta, GA 30339
404-446-1378

Royer Industries, Inc.
PO Box 1233
Kingston, PA 18704
717-287-9624

Tink, Inc.
2361 Durham-Dayton Hwy.
Durham, CA 95938
916-895-0897

MIXING

Brown Bear Corp.
PO Box 148
Lenox, IA 50851
515-333-4551

Can-Am Mixers
PO Box 5318
Fullerton, CA 92635

Hobbs-Adams Engineering Co.
1100 Holland Road
Suffolk, VA 23434
804-539-0231

JWI, Inc.
2155 112th Ave.
Holland, MI 49423
616-772-9011

Knight Industrial Div.
PO Box 167
Brodhead, WI 53520
608-897-2131

Koch Engineering
PO Box 8127
Wichita, KS 67208
316-832-5110

Koppers Co., Inc.
Sprout Waldon Div.
Muncy, PA 17756
717-546-8211

Littleford Bros., Inc.
7451 Empire Drive
Florence, KY 41042
606-525-7600

McLanahan Corp.
200 Wall Street
Hollidaysburg, PA 16648
814-695-9807

Processall
10596 Springfield Pike
Cincinnati, OH 45215
513-771-2266

Scat Engineering
PO Box 266
Delhi, IA 52223
319-922-2230

Scott Equipment
605 4th Ave., N.W.
New Prague, MN 56071

Sludge Systems, Inc.
1125 Starr Ave.
Eau Claire, WI 54702
715-835-6625

ODOR CONTROL SYSTEMS

BioFiltration, Inc.
PO Box 15268
Gainesville, FL 32604
904-375-8001

Calvert Environmental
5985 Santa Fe Street
San Diego, CA 92109
619-272-0050

NuTech Environmental Corporation
5350 N. Washington St.
Denver, CO 80216
800-321-8824

Pepcon Pacific Engineering & Production Co. of Nevada
8201 Gibson Road
PO Box 797
Henderson, NV 89015
702-565-8741

Purafil, Inc.
PO Box 287
Onconomowoc, WI 53066
414-567-0094

Quad Environmental Systems
4284 Overland Trail
Kettering, OH 45429
513-293-8496

SCREENS

Enviro-Care Co.
5614 W. Grand Ave.
Chicago, IL 60639

Falcon Equipment, Inc.
PO Box 339
Gardena, CA 90247
213-327-4880

Hobbs-Adams Engineering
1100 Holland Road
Suffolk, VA 23434
804-539-0231

Krebs Engineers
1205 Chrysler Dr.
Menlo Park, CA 94025
415-325-0751

Lindemann Recycling
500 Fifth Ave.
New York, NY 10110
212-382-0630

Lindig Manufacturing
Box 106
1877 W. County Rd.
St. Paul, MN 55113
612-633-3072

Parker Manufacturing, Inc.
18012 Bothell Highway, S.E.
Bothell, WA 98012
206-486-3547

Powerscreen of America
11300 Electron Drive
Louisville, KY 40299
502-267-2316

Rader Companies, Inc.
PO Box 20128
Portland, OR 97220
503-255-5330

Read Corporation
PO Box 1298
Middlebore, MA 02346
800-992-0145

Resource Recovery Screens
P.O. Box 32035
Detroit, MI 48232
519-977-9852

Royer Industries
PO Box 1232
Kingston, PA 18704
717-287-9624

SPM Group, Inc.
Hwy 16 & 52 N.
Preston, MN 55965
507-765-2107

Sweco, Inc.
PO Box 4151
Los Angeles, CA 90051

Triple/S Dynamics, Inc.
1031 S. Haskell Ave.
Dallas, TX 75223
214-828-8600

SHREDDERS & GRINDERS

American Pulverizer Co.
5540 W. Park Ave.
St. Louis, MO 63110
314-781-6100

Bandit Industries
6750 Millbrook Rd.
Remus, MI 49340
517-561-2270

Blower Application Co.
W19125 Clinton Drive
Germantown, WI 53022
414-255-3446

Columbus McKinnon Corp.
Audubon & Sylvan Pkwys.
Amherst, NY 14228
716-689-5400

Disposable Waste Systems
2600 S. Garnsey
Santa Ana, CA 92707
714-751-8100

Dresser Industries/Jeffrey
PO Box 387
Woodruff, SC 29388
803-476-7523

Eidal Intl., Inc.
PO Box 509
Wilsonville, OR 97070
503-682-2992

Farmhand, Inc.
6421 Hazeltine Blvd.
Excelsior, MN 55331
612-474-1941

Fuel Harvesters Equipment
12759 Loma Rica Dr.
Grass Valley, CA 95945
916-272-7664

Gruendler Crusher
212 S. Oak
Durand, MI 48429
517-288-6662

Hi-Torque Shredder Co.
230 Sherman Ave.
Berkeley Heights, NJ 07922
201-464-2002

Hammermills, Inc.
800 First Ave., N.W.
Cedar Rapids, IA 52405
319-365-0441

Hobbs-Adams Engineering
1100 Holland Road
Suffolk, VA 23434
804-539-0231

Iggesund Recycling
PO Box 380
Nisswa, MN 56468
218-963-4343

Jacobson, Inc.
2445 Nevada Ave., N.
Minneapolis, MN 55427
612-544-8781

Jones Manufacturing Co.
Route 1, Box 80
Beemer, NE 68716
402-528-3861

Komar Industries
4425 Marketing Place
Groveport, OH 43125

La Bounty Manufacturing
PO Box B
Two Harbors, MN 55616
218-834-2123

Lindemann Recycling
500 Fifth Avenue
Suite 1234
New York, NY 10110
212-382-0630

Mac/Saturn Corp.
201 E. Shady Grove Rd.
Grand Prairie, TX 75050
214-790-7800

Carthage Machine Co.
571 W. End Ave.
Carthage, NY 13619
315-493-2380

Montgomery Industries
PO Box 3687
Jacksonville, FL 32206
904-355-5671

Morrison-Knudsen Co., Inc.
PO Box 7808
Boise, ID 83729
800-635-5000

Newell Mfg. Co.
PO Box 9367
San Antonio, TX 78204
512-227-9090

O&E Shredding Systems
PO Box 69
Green Bay, WI 54305
414-437-6587

Parker Manufacturing
18012 Bothell Hwy., S.E.
Bothell, WA 98012
206-486-3547

Proveda
100 Jerry Drive
Jackson Center, OH 45334
513-596-6713

Recycling Systems
PO Box 364
Winn, MI 48896
517-866-2800

Shred-Pax Limited
PO Box 1508
Cambridge, Ontario
Canada

Shredding Systems, Inc.
PO Box 869
Wilsonville, OR 97070
503-682-3633

Stumpmaster, Inc.
PO Box 103
Rising Fawn, GA 30738
404-462-2445

Triple/S Dynamics, Inc.
1031 S. Haskell Ave.
Dallas, TX 75223
214-828-8600

United Farm Tools, Inc.
PO Box 336
Turlock, CA 95381
209-632-3846

Van Dyk Baler Corp.
234 Fifth Ave.
New York, NY 10001
212-683-2610

Williams Patent Crusher & Pulverizer Co.
2701 N. Broadway
St. Louis, MO 63102
313-621-3348

R.M. Wilson Co.
Wheeling, WV
800-624-5460
800-852-5411 (WV only)

Wor-Tex Corp.
PO Box 636
Hillsboro, TX 76645
817-582-5354

SPREADERS & APPLICATORS

Ag-Chem Equipment Co.
4900 Viking Drive
Minneapolis, MN 55435
612-835-2476

Field Gymmy
Box 121
Glandorf, OH 45848
419-538-6511

Floater Vehicle, Inc.
PO Box 24
Hallsville, OH 45633
614-655-2095

Henderson Mfg.
PO Box 40
Manchester, IA 52057
319-927-2828

Hobbs-Adams Engineering
1100 Holland Road
Suffolk, VA 23434
804-539-0231

I.M.E.
PO Box N
Galva, IL 61434
309-932-2036

Knight Mfg.
1501 W. 7th Ave.
Brodhead, WI 53520
608-897-2131

L.W.T., Inc.
Liquid Waste Technology
Box 250
422 Mill Street
Somerset, NJ 54025
715-247-5464

New Holland
500 Diller Ave.
New Holland, PA 17557
717-354-1121

S & ME
PO Box 1308
Cary, NC 27511
919-481-0397

Stallon-Beiler Equipment Company, Inc.
R.D. 1
Massey, MD 21650
301-928-3009

TUBING, AERATION

Advanced Drainage Systems
PO Box 5807
Columbus, OH 43221
1-800-848-8195

Hancor, Inc.
PO Box 1047
Findlay, OH 45839
1-800-2-Hancor

WINDROW TURNING EQUIPMENT

Brown Bear
P.O. Box 148
Lenox, IA 30851
515-333-4551

Cobey Composting
4250 S.R. 309
Gailon, OH 44833
419-468-2288

Resource Recovery Systems of Nebraska, Inc.
Route 4
Sterling, CO 80751
303-522-0663

Scarab Mfg. and Leasing, Inc.
Route 2
White Deer, TX 79097
806-883-7621

Scat Engineering
P.O. Box 265
Delhi, IA 52223
319-922-2230

Wildcat Mfg. Co., Inc.
P.O. Box 523
Freeman, SD 57029
605-925-4512

Notes

Notes

Notes

Notes